PASSIVE SOLAR HEATING DESIGN

PASSIVE SOLAR HEATING DESIGN

RALPH M. LEBENS

B.A., B.Arch.(Manch.), M.Arch.A.S.(MIT), R.I.B.A.

Consultant, Architecture and Computer-Aided Energy Design, London, UK
Formerly Research Assistant to the MIT Solar Building 5 Project,
Massachusetts Institute of Technology, Cambridge, Mass., USA

APPLIED SCIENCE PUBLISHERS LTD
LONDON

APPLIED SCIENCE PUBLISHERS LTD
RIPPLE ROAD, BARKING, ESSEX, ENGLAND

First edition 1980
Reprinted 1981

British Library Cataloguing in Publication Data

Lebens, Ralph M.
 Passive solar heating design.
 1. Solar heating—Passive systems
 I. Title
 697'.78   TH7413

ISBN 0-85334-870-7

WITH 46 TABLES AND 77 ILLUSTRATIONS

© APPLIED SCIENCE PUBLISHERS LTD 1980

Printed in Great Britain by Galliard (Printers) Ltd, Great Yarmouth

With thanks to Simone
and to my teachers: my parents,
Kultar Singh, and Timothy Johnson

Preface

This book was written in its initial form at the Massachusetts Institute of Technology during the design and construction stages of the MIT Solar 5 building in 1977–78. It has been revised and extended to provide a working handbook for building designers and researchers working within the field of passive solar heating design.

A brief review of the various types of passive solar systems is provided in Chapter 2. Despite the growing publicity given to passive solar design and the apparent success and cost-effectiveness of such an approach, the design concepts are not being widely adopted yet. This is largely due to the lack of simple design methods available. Although the calculations of heat losses and solar gains are well understood and documented, the quantitative analysis of storage and control of the solar gains is not yet within the grasp of the building designer. Nor does he have mastery of predicting the cost-effectiveness of implementing a passive design improvement at design stage. Such a prediction will have to be made for almost every passive solar project until the design approaches gain credibility and have been accepted within the hard core of the architectural profession and the construction industry. The aim of the research behind this book was to remedy this situation by providing a workbook of passive solar design methods for building designers. Another obstacle in the path of widespread architectural use of passive concepts is that the analytical design methods required would be complex and long-winded, requiring either too much time for the tight schedule of an architectural practice, or the use of a computer. Few practices have access to a computer, so the design methods are presented in the form of programs for the TI–59 (Texas Instruments programmable desk-top calculator). This instrument is within the budget of a one-man architectural practice and the programs take only minutes to run.

An example design and analysis is provided as an illustration of some of the methods presented in this book.

A deliberate attempt is made to demystify the problems by avoiding useless complications such as the use of joules or gigajoules as measurements of energy rather than familiar units of watt-hours or kilowatt-hours. All units are provided in both metric and imperial.

RALPH M. LEBENS

Contents

Nomenclature

A = Area (m², ft²)

A_c = Area of collection window (m², ft²)

A_{ch} = Number of air changes per hour

A_d = Hourly area of window in potential direct sunshine (m², ft²)

A_f = Hourly area of window in shade (m², ft²)

A_{ws} = Area of the weather skin (m², ft²)

$a = (1 + g)/(1 + r)$

B_{ni} = The information supplied by matrix **B**

C = Infiltration coefficient (m³/h m N/m², ft³/h ft lbf/ft²)

C_n = The information supplied by matrix **C**

Cap = Capacitance (Wh/°C, Btu/°F)

Con = Conductance (W/°C, Btu/h °F)

C_l = Correction factor defined by eqn (2), Section 3.5 (fraction)

DD = Degree days per month

D = Dimension considered (m, ft)

dT = Internal → external temperature difference (1 °C, 1 °F)

F = Current year's value of the annual fuel savings (£, $)

f_n = A factor representing the proportion of total incoming solar energy falling on node 'n' (fraction)

g = 'Real' rate of fuel prices increase (fraction)

H = Number of hours in the day (= 24)

I = Total clear day incident solar energy on the collection surface (Wh/m², Btu/ft²)

I_{dh} = Hourly direct insolation on the horizontal surface (W/m², Btu/h ft²)

I_{dn} = Hourly direct insolation on a surface normal to the solar radiation (W/m², Btu/h ft²)

I_{dv} = Hourly direct insolation on a vertical surface (W/m², Btu/h ft²)

I_{fh} = Hourly diffuse insolation on a horizontal surface (W/m², Btu/h ft²)

I_h = Monthly total insolation on a horizontal surface (W/m², Btu/h ft²)

I_{hor} = Hourly horizontal component of the insolation value (W/m², Btu/h ft²)

I_{max} = Maximum hourly insolation per unit area on the surface and month considered (W/m^2, Btu/h ft^2)

I_s = Hourly total insolation per unit area on the surface and month considered (W/m^2, Btu/h ft^2)

I_t = Monthly total radiation transmitted through a unit area of a vertical south-facing double-glazed window (W/m^2, Btu/h ft^2)

I_{tav} = Mean monthly total insolation on the surface considered (Wh, Btu)

I_{tcl} = Monthly clear day total insolation on the surface considered (a month of clear days) (Wh, Btu)

I_{tot} = Daily total insolation on a given surface (W/m^2, Btu/h ft^2)

I_{tv} = Hourly total insolation on a vertical surface (W/m^2, Btu/h ft^2)

I_{tvdf} = Hourly total clear day insolation for the vertical orientation considered (W/m^2, Btu/h ft^2)

I_{tvf} = Hourly diffuse insolation on the vertical surface considered (W/m^2, Btu/h ft^2)

L = Latitude (deg.)

l = Gap length (m, ft)

M = Correction factor for area of frames (0·85) times a dirt factor (fraction)

M_n = Month number

NPV = Net present value of the money saved over N years (£, $)

N_d = Number of days per month

N_y = Lifetime of improvement or number of years of borrowing (y)

P = Heat capacity of air (specific heat × density) (Wh/m^3 °C, Btu/ft^3 °F)

Q_g = Net gain of solar energy for a clear day (Wh, Btu)

Q_h = Hourly horizontal component of the solar gains on a clear day (Wh, Btu)

Q_i = Incidental energy gains per day from people, appliances and solar energy (Wh, Btu)

Q_l = Heat loss per day per degree (Wh/day°C, Btu/day °F)

Q_{lh} = Heat loss per hour through the collection window $U × A × \Delta T$ (W, Btu/h)

Q_m = Monthly heating load (Wh, Btu)

Q_{sg} = Hourly solar gain for the window concerned (W, Btu/h)

Q_{st} = Fraction of the absorbed energy which is stored in the thermal mass

R = Reflection (fraction)

R_{air} = Resistance to energy flow away from mass into the air (m^2 °C/W, ft^2 h °F/Btu)

R_l = A correction factor for back-reflected losses (fraction)

R_{si} = Internal surface resistance (m² °C/W, ft² h °F/Btu)
R_{slab} = Resistance to energy flow into the slab (m² °C/W, ft² h °F/Btu)
R_{so} = External surface resistance (m² °C/W, ft² h °F/Btu)
r = 'Real' discount rate (fraction)
r = Radius of curvature of slat (mm, in.)
S = Slat width (mm, in.)
s = Separation of slats (mm, in.)
T = Temperature (°C, °F)
T_A = Outdoor temperature (°C, °F)
T_{DD} = Balance point or degree day base temperature (°C, °F)
T_a = Ambient temperature for the hour being referred to (°C, °F)
T_{ad} = Hourly direct insolation heat gain factor for the material considered (fraction)
T_{adf} = Average heat gain factor for that month, orientation and glazing material (fraction)
T_{af} = Diffuse insolation gain heat factor (fraction)
T_{av} = The normal average diurnal temperature swing for that month
T_d = Average diurnal outdoor temperature (°C, °F)
T_o = Mean monthly outdoor temperature (°C, °F)
T_{sw} = The normal diurnal temperature swing for that month (°C, °F)
T_{th} = Average thermostat set point temperature (°C, °F)
t = Time
t_{sr} = Time of sunrise
U = U-value (W/m² °C, Btu/h ft² °F)
V = Volume of air in the building (m³, ft³)
V_a = Volume of air flow (m³/h, ft³/h)
v = Velocity (m/s, ft/s)
W = Sun–wall azimuth angle (deg.)
ΔT = Temperature difference (°C, °F)
Δp = Pressure difference across the window (half the pressure difference across the building) (N, lbf)
Δt = Time difference (h)
α = Absorption (of the node) (fraction)
γ = Viscosity (m²/s, ft²/s)
δ = Mid-month solar declination (deg.)
ε = Emission (fraction)
η = Efficiency (fraction) of the system
θ = Angle of incidence (deg.)
τ = Transmission (fraction)
ψ = Angle of solar altitude (deg.)

Chapter 1

Active versus Passive—Towards a Design Philosophy

An outline of the concepts involved in solar design is needed to identify the reasons for the preliminary design standpoint taken in this book. There are two approaches to low-energy house design. The first uses solar collecting panels, storage tanks or bins, an energy transfer mechanism and an energy distribution system. It is known as an active system (Fig. 1.1) and always employs one or more working fluids which collect, transfer, store and distribute the collected solar energy. The working fluids are circulated by means of fans or pumps. Since they are usually photogenic, houses designed with this approach receive much publicity in the architectural press.

The second approach, passive design (Fig. 1.2), seeks to reduce the house's energy budget by close attention to orientation, insulation, window placement and design, and to the subtleties of the energy transfer properties of building materials. Since solar gains are present in every building, all buildings are passively solar-heated to some extent. It is when the building has been designed to optimize the use of solar energy and when solar energy contributes substantially to the heating requirements of the building that it is termed a solar building.

It is only in an era of cheap fuel that we have lost the skills of designing with the climate. Practised in all indigenous architecture, it is an art we shall have to relearn. As told by the variety of form of vernacular architecture, such design considerations will allow the building to respond to its location. Much like a plant, which in a dark corner will grow towards the light, the form of a passive solar building will vary with change in site and even more with change in climate. The passive solar approach presents a comprehensive design philosophy.

Conceptual comparison of passive and active designs is made by considering the passively designed house as acting as absorber, store and delivery—there is usually no working fluid: all the functions of the active systems are carried out by the building materials themselves. Most examples of passively solar-heated buildings use extensive south-facing glazing areas to admit low-angle winter sunshine into the building, and

1

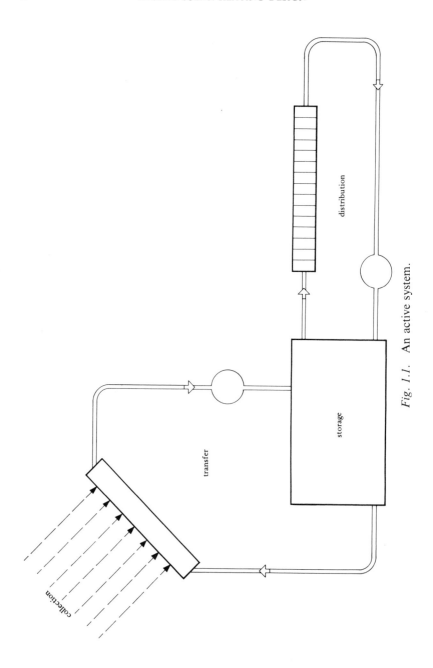

Fig. 1.1. An active system.

extensive mass heat capacity inside the thermal envelope of the building to store the admitted energy. The cost of heating (and cooling) is then the increased cost of the materials required for passive design above that of the conventional building materials they replace.

Although often less spectacular and graphic, this second approach produces better results in terms of energy conservation and money saved as

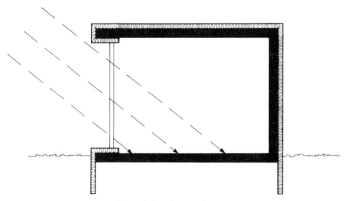

Fig. 1.2.   A passive system.

a result of initial expenditure. The method used for cost-benefit comparisons is termed Life-cycle Costing (LCC). This small area in the field of solar energy collection is one of the keys to successful solar design; for the aim is always comfort and economic feasibility, not 100% solar heating. The present value of the money saved over n years is computed with corrections for the effects of the borrowing interest rate and the rate of increase in the price of fuels.* From this it is possible to calculate the number of years it will take to recoup the capital cost of an improvement. An active space heating system will have difficulty in producing less than a 30-year payback figure. With passive systems, payback periods of anything from 1 to 13 years are usual. The durability of passive systems further increases their economy. They require little maintenance and will last the lifetime of the building. Thus they will not require replacement, as may be the case with certain parts of active systems.

Under overcast skies or intermittent sunshine an active system will shut down but a passive system continues collecting. In many climates this diffuse solar energy is not insignificant [1].

* See Section 3.18 ('Cost Analysis'), for a more detailed account of Life-cycle Costing.

A combination of the two systems is often sought if passive design will not meet a large enough portion of the total heating requirements. The feasibility of this remains subject to the economic analysis of each case.

Although passive design seems to hold the answers it does have some large design problems. Access, space use, view and ventilation conflict with the energy collection requirements, and the problems of overheating and possibly glare must be overcome to provide comfort within the passive solar building. These are some of the design parameters dealt with in depth in the following chapters.

REFERENCE

1. COURTNEY, R. G., *Solar Energy Utilization in the UK: Current Research and Future Prospects*, Building Research Establishment CP/64/76, October 1976.

Chapter 2

Classification and Comparison of Passive Systems

2.1. CLASSIFICATION

The two predominant approaches to passive solar heating are indirect and direct gain systems. All other categories are subsets within these two approaches.

The following subsets are included within the indirect heat gain category:

(a) Trombe–Michel wall
(b) Water wall
(c) Roof pond
(d) Attached/isolated sunspace (or greenhouse)
(e) Thermosyphoning collectors (air or water)

In contrast there are only two types of direct heat gain systems:

(a) Diffusing direct gain
(b) Non-diffusing direct gain

To simplify a comparison of the difference between the concepts involved in indirect and direct heat gain systems, the Trombe–Michel wall and the non-diffusing direct gain systems are compared.

The indirect approach employs external building elements to absorb the solar radiation (Fig. 2.1). In the Trombe–Michel wall example, these building elements are massive and have no thermal barrier so that the energy absorbed travels by conduction to the internal surface. It takes several hours for this to happen, at which point it provides the room with radiant and convective heat. Movable insulation and cover glazing may be externally applied to the absorbing surfaces, depending on the climate. If cover glazing has been applied, room air can be circulated across the absorbing surface to provide instant heating.

The direct heating approach uses a transparent wall to allow solar radiation to enter the space requiring heating (Fig. 2.2). The energy is absorbed and stored by the areas of thermal mass* it strikes. The thermal

---

* Thermal mass describes a building element (floor, wall or ceiling) with a high thermal capacity.

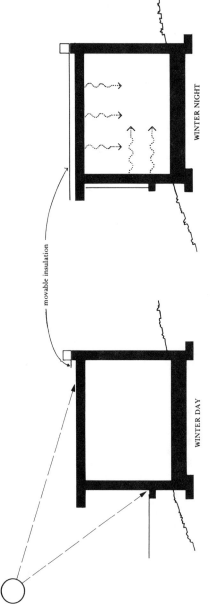

Fig. 2.1. Indirect heat transfer to interior [1].

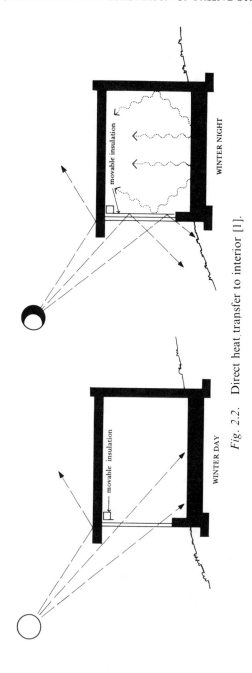

Fig. 2.2. Direct heat transfer to interior [1].

mass is isolated from the external climate by an insulating envelope. The thermal mass then heats the room by radiation and convection. Movable insulation may be applied to the transparent wall to prevent excessive heat losses at night.

The following brief outline of examples of both approaches will give some insight into the design conflicts inherent in passively heated buildings.

2.2. INDIRECT GAIN EXAMPLES

(a) Trombe–Michel Wall
The original examples of this passive type are found in Odeillo in southern France (Figs. 2.3, 2.4). They were designed by Dr Felix Trombe of the Centre National de la Recherche Scientifique (CNRS), and an architect, Jacques Michel.

The Odeillo houses (1967) use a black-painted 600-mm (24-in) thick concrete south wall to collect and store energy from the low-altitude winter sun. Heat losses are reduced by the use of double glazing across the face of this wall. The primary method of heating is by convection from the inside surface of this massive concrete wall to the room. It takes 10 to 15 hours for the heat to travel through the wall so instantaneous heating is made possible by circulating room air in the gap between the glass and wall. This is done without the use of fans; dampers are used on the lower openings to shut off thermosyphoning at night. The remainder of the structure of this two-bedroomed house-type is well insulated so that the heat requirements are

Fig. 2.3.   The Odeillo house-type (1967 prototype), France.

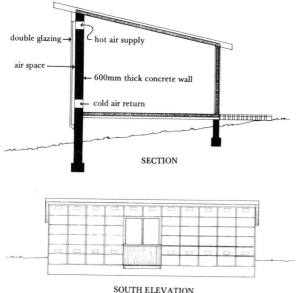

double glazing →   hot air supply

air space ──

← 600mm thick concrete wall

← cold air return

SECTION

SOUTH ELEVATION

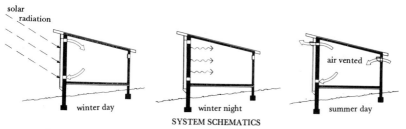

solar
radiation

air vented

winter day   winter night   summer day
SYSTEM SCHEMATICS

Fig. 2.4.   The Odeillo house-type (1967 prototype), France [3].

low. The overhang of the roof shades the south wall completely during the summer and the glazing vents can be opened to encourage cross ventilation.

The results achieved in this instance represent 65–70 % of the annual heating load of the house. The following data refer to the 1967 prototype houses at Odeillo [2]:

Latitude: 42°29′
Global radiation on a horizontal surface: 1605 kWh/m² y (547 600 Btu/ft² y)
Hours of sunshine: 2500/y
Heat loss of building: 11·736 kW/day °C (22 250 Btu/day °F)

Collection area: 48 m² (480 ft²)
Floor area: 76 m² (760 ft²)
Air gap between wall and glass: 120 mm (4¾ in)
Dimension of each panel (not self-contained): 4·38 m (14 ft 6 in) high
by 1·27 m (4 ft 3 in) wide
Distance between ducts: 3·5 m (11 ft 8 in)
Dimension of ducts: 565 mm (1 ft 10½ in) wide by 110 mm (4½ in) high

Indoor room-air overheating after clear days in spring and autumn has been reported.

Several design conflicts are noticeable in this house-type. The south wall is unwillingly punctured by a door. The provision of access or view is possible only at the expense of the solar system. North windows are provided for additional daylighting; these will add little light, increase heat losses and reduce the mean radiant temperature within the room. The result of this

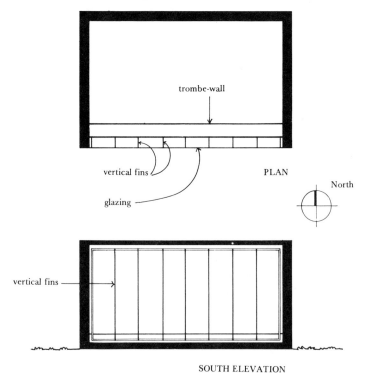

Fig. 2.5. The 'Peakshaver' solar collector by the Environmental Research Laboratory, University of Arizona.

overriding concept for solar heating is a building with little flexibility, but one which has its appropriate applications.

There are some new ingenious ideas which have been independently developed and which go a long way to reducing some of the disadvantages of the Trombe–Michel system. The Environmental Research Laboratory of the University of Arizona [4] proposed the use of vertical fins attached to the wall so that their east side is painted black and their west side is reflective (Fig. 2.5). Morning insolation strikes the blackened side of the fins and hot air is quickly supplied to the room by convection. Thus the room is heated long before the Trombe–Michel wall is. In the afternoon the reflective side of the fins is seen by the sun and therefore all the energy is directed on to the wall.

James Bier has built a house in Ferrum, Virginia, USA [5] in which he has divided the conventional Trombe–Michel wall into five short lengths, each turned so that it is at an angle of 45° to the south glazing (Fig. 2.6). The

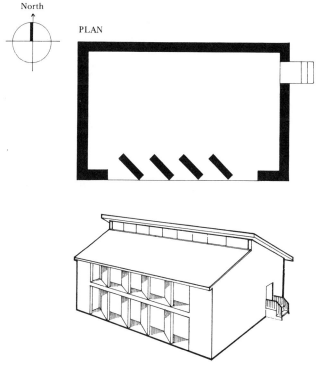

Fig. 2.6.   James Bier's house in Virginia, USA.

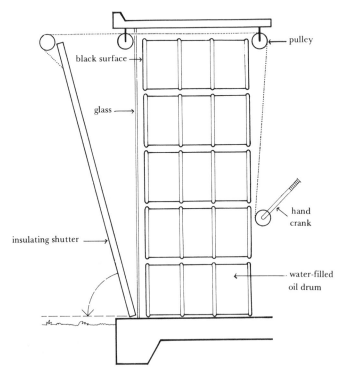

SECTION THROUGH COLLECTION WALL

Fig. 2.7. Steve Baer's house in Corrales, Albuquerque, New Mexico, USA (1971) [6]. (Photograph by courtesy of New Mexico Solar Energy Association.)

concrete walls are each 800 mm (2 ft 4 in) by 300 mm (1 ft) and at 1·2 m (4 ft) centres. This has become a mixed passive system consisting of half direct gain and half indirect gain. By so doing it has reduced many of the architectural disadvantages of the Trombe–Michel wall: it allows morning sunlight in, it intercepts and stores afternoon sunlight thus preventing overheating, and it allows view and access through the south collecting surface.

(b) Water Wall

The water wall is similar in concept to the Trombe–Michel wall but uses contained water instead of masonry. Water heat storage is of great potential benefit because it distributes heat gains quickly by convection and thus it has the capability of providing passive solar energy storage with greatly reduced surface temperatures. This will reduce the possibilities of spring and autumn overheating within the space. Water is also attractive because it is cheap, easy to install and compared with concrete has a high thermal capacity per unit volume. The problems with water are those of containment, corrosion and flood risk.

The first example of a water wall system was Steve Baer's house in Corrales, New Mexico, USA (see Fig. 2.7). This was built in 1971 and uses 56-gallon steel barrels stacked behind single-glazed south-facing windows which are covered during winter nights by fold-down insulating panels. These panels also act as reflectors in winter but the operation reverses in summer to provide night-time radiative cooling and day-time insulation and shading from the sun. The flow of heat from the barrels to the rooms is controlled in winter by curtains.

(c) Roof Pond

This system was invented by Harold Hay. The first such system in the form of a dwelling was built in 1973 by him and Kenneth Haggard at Atascadero, California, USA (Figs 2.8–2.10). It employs containerized water in the roof for its thermal mass. The containers are four 2·7 m (8 ft) by 12·7 m (38 ft) by 200 mm (8 in) deep PVC plastic bags. These are supported by steel decking which is also the ceiling. The water-filled bags are covered by movable insulation which retracts to allow solar radiation to be collected during the winter and radiational cooling in summer to the night sky. The house below is heated by radiation from the steel ceiling. The structural walls needed to support this mass of water also serve the purpose of secondary thermal mass. This is a single-storey house with a floor area of 107 m² (1140 ft²). The

Fig. 2.8. The Atascadero 'Skytherm' house (1973), California, USA. Photograph
by Harold Hay.

system gives good thermal comfort conditions; no auxiliary heating has
been needed since it was built. The insulation panels are moved by a $\frac{1}{4}$ kW
($\frac{1}{3}$ hp) motor. The limitations of this method of solar heating are less easily
detected. The structural walls needed reduce the flexibility of the
permissible spaces. The method is limited to a single-storey structure and to
areas where there is little snowfall. Nevertheless, this system is extremely
well adapted to its particular location, because it can be used to either cool
or heat the dwelling in a climate where both are of equal importance.

Fig. 2.9. The water-filled PVC bags on the roof of the 'Skytherm' house. The PVC
is transparent although it appears to be opaque in this photograph (by Harold Hay).

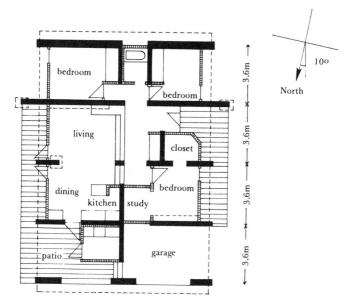

Fig. 2.10. The Atascadero 'Skytherm' house. Plan [7].

Fig. 2.11. Unit 1, First Village, Santa Fe, New Mexico, USA (1975).

FLOOR 2

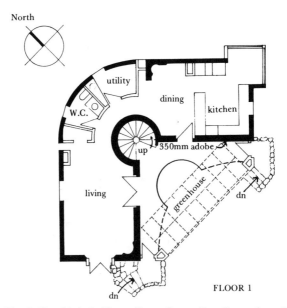

FLOOR 1

Fig. 2.12.   Unit 1, First village, Santa Fe—floor plans [7].

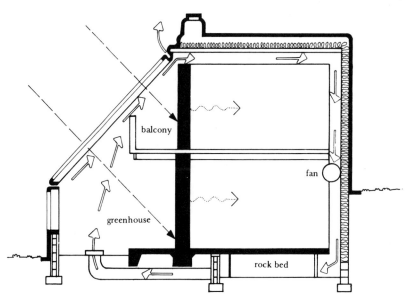

Fig. 2.13. Unit 1, First village, Santa Fe—section [7].

(*d*) *Attached/Isolated Sunspace* (*or Greenhouse*)

Probably the best known example of this type of system is Unit 1 of First Village, Santa Fe, New Mexico, USA (Figs 2.11–2.13). Designed and built in 1975 by Wayne and Susan Nichols it is the house in which Doug and Sara Balcomb live. This type of system is usually linked to storage within the house by means of forced air circulation and therefore termed a 'hybrid'. An isolated sunspace in itself will do little more than halve the yearly heat losses from the wall to which it is attached. There will be times when the sunspace overheats and unless this energy is stored for later use it will be wasted. Storing this energy usually involves the use of a fan, attached to a temperature differential switching device, to circulate the hot air to storage. In Unit 1, Santa Fe, the storage consists of two rock beds in the ground floor, one in each wing. The solar benefit of this type of passive system cannot be separated from the benefit of growing food within the sunspace. Calculating gains from solar energy only is made difficult because hourly simulation is needed to assess the quantity and timing of excess solar heat storage.

(*e*) *Thermosyphon System*

The most common type of thermosyphoning system is for passive solar

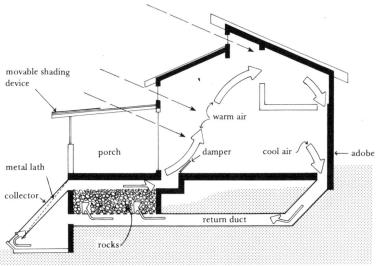

Fig. 2.14. The Paul Davis house, Corrales, Albuquerque, New Mexico, USA (1972) [6]. (Photograph by courtesy of New Mexico Solar Energy Association.)

domestic water heating. There are only a few examples of thermosyphoning collectors being used for space heating, the most publicized of which is the Paul Davis house, Albuquerque, New Mexico, USA (Fig. 2.14). This was built in 1972 and is 75 % solar heated. This house has a floor area of 100 m² (1000 ft²) and a single-glazed air collector area of 42 m² (420 ft²). The house is also heated by direct gain. Forty-five tons of rock storage is placed beneath the front porch and dampers control the flow of solar-heated air into the house.

The flow of air through a thermosyphoning collector (Fig. 2.15) is driven by the difference in weight per unit volume between the unheated and heated columns of air. This is a function of the difference in average air temperatures between the two columns, and their height. Therefore

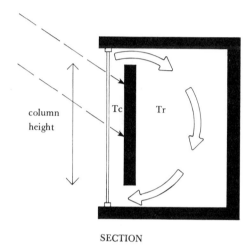

SECTION

Fig. 2.15.   A typical thermosyphon loop.

optimizing the collector means increasing the temperature difference between the two columns of air and increasing the column heights. A third factor influences air flow rates, and thus the collection efficiency: the cross-sectional area(s) of the flow channel(s) within the collector and of the inlet and outlet vents. All vents should, if possible, have the same area as that of the channel (i.e. unrestricted flow), although if this area is halved there is only a 10% reduction in total heat benefit from the collector. The cross-sectional area of the channel within the collector should be at least $\frac{1}{20}$ of the collector area. A maximum channel area is about $\frac{1}{10}$ of the collector area; an increase above this may decrease collector performance by setting up convection currents within this space.

The majority of research work on passive thermosyphoning systems has been conducted by the Centre National de la Recherche Scientifique (CNRS) (France) and by W. Scott Morris (USA). There seems to be a conflict of opinion between these two researchers. According to Scott

Morris, if there is a temperature difference of 27·5–36 °C (50–65 °F) the flow is acceptable and efficiency is reasonable. He also describes the usual thermosyphon flow as being 180 mm/s (0·6 ft/s) [8]—this happens to be the design flow rate for an active system also. The thermosyphon space heating panels on the façade of the CNRS office building in Odeillo (Fig. 2.16) did not work originally. This was because the separating baffle was made of thin

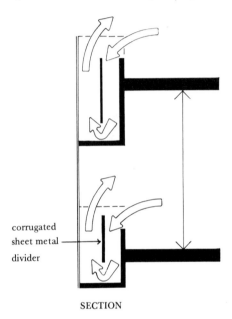

corrugated
sheet metal
divider

SECTION

Fig. 2.16. The thermosyphon panel heaters on the CNRS building at Odeillo, France.

corrugated steel which transmitted the solar heat gain to both air columns with almost equal speed. The resulting lack of temperature difference between the two air columns prevented free thermosyphoning. Improvements were made by insulating this separating baffle and an air velocity of 300 mm/s (1 ft/s) was achieved. The maximum temperature difference between inlet and outlet was 30 °C (54 °F) which, according to Scott Morris, should give reasonable collection efficiencies. The collection efficiencies at the CNRS were found to have a maximum value of 18 % and that for the heating season was 13 % of the incident solar energy. These efficiencies are for the improved thermosyphoning collector and cannot be considered to be reasonably good.

2.3. DIRECT GAIN EXAMPLES

(a) Diffusing Direct Gain System

The forerunner of the examples of this type of passive system is the two-storey 77 m (230 ft) by 12 m (37 ft) extension to St George's Secondary School, Wallasey, England (Fig. 2.17). The designer was E. A. Morgan and the building was completed in 1961. The entire south wall of 500 m²

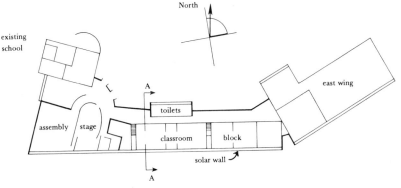

PLAN

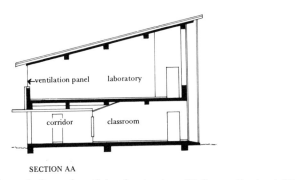

SECTION AA

Fig. 2.17.   St George's Secondary School extension, Wallasey, England [9].

(5350 ft^2) is double glazed with the panes set 600 mm (2 ft) apart to permit easy cleaning. The 180-mm (7-in) thick concrete roof and 230-mm (9-in) thick brick wall form the thermal mass and are insulated externally with polystyrene. Twelve percent of the area of the inner glazing skin on the south wall is black-painted masonry. Thirty-three percent of the south wall area has reversible aluminium panels which are hung bright side facing outward from April to October and black side facing outward from November to March. Almost all the inner glazing skin is diffusing glass to prevent discomfort glare from direct sunshine. There have been no reported complaints of glare. Ventilation, by means of openable dampers, and auxiliary heating, by switching on the lights, are the only forms of thermal control. The building has functioned over the past 18 years without any auxiliary heating. The average outdoor temperature in winter is 4·5 °C (40 °F) with a 9 °C (16 °F) fluctuation. On average the indoor temperature fluctuation was found to be one half of the outdoor temperature fluctuation; room temperature fluctuation was usually 5 °C (9 °F) on a clear winter day. The design limitations imposed by such a system are potential glare caused by such a large area of glass, the conflict between preventing this and still allowing view, and the conflict between space use and energy absorption. The low ventilation levels needed to maintain comfortable temperatures have caused occasional complaints of body odours. Admittedly, most of these conflicts have been acceptably resolved in this example but the use of this system for a dwelling would pose the same problems without such readily available solutions to hand. The final design conflict is the enormous structural mass required. If this were a block of apartments such mass requirements would cause greater structural and foundation costs than those normally encountered.

(b) Non-diffusing Direct Gain System
An example of this system is the MIT Solar 5 demonstration building designed by Timothy E. Johnson. This building was completed in February 1978 (Fig. 2.18). It is used as an example here because it illustrates an approach which, using newly developed materials, considerably reduces most of the design limitations we have noted so far. There are two exceptions to this: the system must be used in the latitudes where winter sun altitudes are low enough to strike the vertical walls within a 50° angle of incidence, and the collection aperture must face due south. This single-storey building has a floor area of 76 m^2 (817 ft^2). The south wall has 16·7 m^2 (180 ft^2) of glazing material. A heat mirror (which reflects infrared radiation without appreciably reducing transmittance to solar radiation) has been

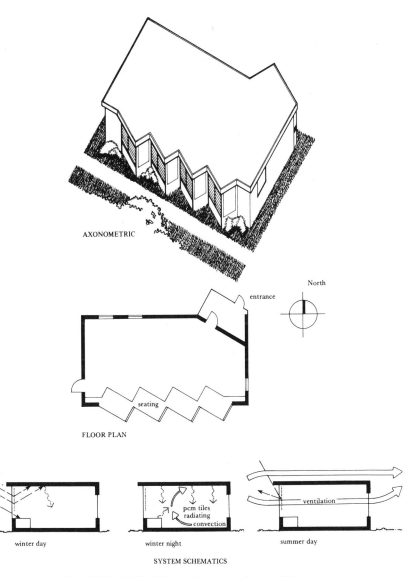

AXONOMETRIC

North

entrance

FLOOR PLAN

seating

radiation

winter day

pcm tiles
radiating
convection

winter night

ventilation

summer day

SYSTEM SCHEMATICS

Fig. 2.18.   MIT Solar 5 demonstration building.

developed by Suntek Inc., California, USA, and is used on the south glazing of this building. Johnson developed a phase change material tile (Sol–Ar–Tile® by Architectural Research Corporation, Livonia, Michigan, USA) which is used as the thermal mass. The tile is 31 mm ($1\frac{1}{4}$ in) thick, coloured blue, and is both used as a ceiling finish and placed on the seating areas below the south windows. Direct solar radiation is reflected up on to the Sol–Ar–Tiles® and stored there in the form of latent heat. The reflective blinds or solar modulators are designed to require a minimum number of (11) adjustments throughout the heating season. The indoor temperature fluctuation on a sunny winter day is approximately 5·5 °C (10 °F). Auxiliary heating is reduced to only 12 % of the total seasonal heating requirements. There is no potential conflict between south glazing and view, access or ventilation. Glare is greatly reduced by the blinds which obscure most of the glare source and increase the surround illumination levels at the back of the room. Discomfort from direct solar radiation is avoided by the design requirement of the reflection angle off the louvres being not less than 30 °. The Sol–Ar–Tiles® weigh a fraction of the weight of an equivalent thermal capacity in other materials, thus no heavy loads are imposed as a solar heating requirement. Because the energy is stored with minimal increase in the surface temperature of the tile and because the phase change is designed to occur at room temperature, the air temperature fluctuations within the building are minimized. Finally, the restrictions on the size and flexibility of the space due to solar-heating requirements are greatly reduced. The components of this system will be discussed in greater depth in the following chapter.

2.4. HYBRID SYSTEMS

Such systems include either a mechanically operated heat storage system or a mechanically operated heat distribution system. The line drawn between passive and hybrid is difficult to define and is of academic value only.

2.5. COMPARISON

It is too early in the field of passive solar architecture to attempt any comparison, in terms of economics and efficiency, using actual buildings. It is suggested that this type of work requires attention but not through the use

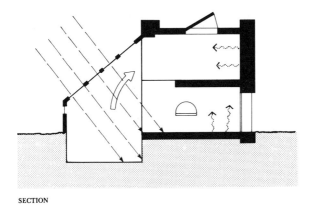

SECTION

Fig. 2.19. An example of linked sunspace and direct gain passive systems—Chino Valley Solar Adobe Studio, Arizona, USA. By Otwell and Frerking [6].

of simulation models as these will not touch on the user interaction with the building, an aspect which is inseparable from passive solar systems.

A general comparative overview can be attempted by considering various factors. The main concern for the efficiency of a passive solar system is that the incoming solar energy is stored as soon as possible and that it is stored at low temperatures. If the energy is not stored directly the usual policy is to allow the collection space air temperature to rise and at a preset temperature the air will be circulated through a storage mass. By not storing the energy immediately there are higher heat losses to the outside—heat loss being proportional to the temperature differential—and therefore lower collection efficiencies. Again high temperature storage (higher than room air temperature) will cause greater heat losses than low temperature storage, and may cause discomfort from overheating of the space. Such overheating will result in the opening of windows and thus wastage of the potentially collectable energy. These observations indicate several simple guidelines: direct gain has a good chance of being more efficient than the Trombe–Michel wall (storage temperatures are lower); for the same reason the water wall or roof pond will be more efficient than the Trombe–Michel wall, and the isolated sunspace has a good chance of being less efficient than any other system. If direct gain or Trombe–Michel systems are employed in conjunction with the isolated sunspace system (Fig. 2.19), the inefficiencies of this system are greatly reduced. It should be realised that efficiency is not the most important factor in passive solar design. For instance the sunspace and direct gain systems have a good chance of providing more interesting and nurturing architectural design solutions. The work of David Wright (California, USA) and Michel Gerber (Perpignan, France) illustrates this point.

REFERENCES

1. JOHNSON, T. E., 'Performance of passively heated buildings', *Journal of Architectural Education*, February 1977.
2. TROMBE, F. *et al.*, 'Some performance characteristics of the CNRS Solar House Collectors', *Passive Solar Heating and Cooling Conference and Workshop Proc.*, ERDA, LA–6637–C, Albuquerque, New Mexico, USA, May 18–19, 1976.
3. ARIZONA STATE UNIVERSITY, *Solar-Oriented Architecture*, 1975. For the AIA Research Corporation and the National Bureau of Standards.
4. THE ENVIRONMENTAL RESEARCH LABORATORY, UNIVERSITY OF ARIZONA, 'The peakshaver solar collector', *2nd National Passive Solar Conference Proc.*, Philadelphia, PA, USA, March 1978, Vol. III, p. 910.
5. IBID. BIER, JIM, 'Vertical solar louvers: a system for tempering and storing solar energy', Vol. I, p. 209.
6. AIA RESEARCH CORPORATION, *A Survey of Passive Solar Buildings*, February 1978.
7. SANDIA LABORATORIES, *Passive Solar Buildings: A Compilation of Data and Results*, June 1978.
8. SCOTT MORRIS, W., 'Natural convection collectors', *Solar Age*, September 1978.
9. SZOKOLAY, S. V., *Solar Energy and Building*, 1975, The Architectural Press and Halsted Press.

Chapter 3

A Workbook of Passive Solar Design Tools and Logic

The aim of the passive design philosophy is always to tame the annual and diurnal fluctuations in temperatures to within comfort standards. This task is far from simple; it consists of many different facets. This chapter provides an outline, a brief background, and a workbook of these facets.

3.1. CLIMATE

Close observation of the winter weather patterns of an area will help in the assessment of the optimum orientation for the building.

Two sets of climatic data are needed to conduct analyses of the passive systems. The first is the 'normals' of daily temperatures for each month of the year. These are required to calculate monthly heating degree day figures.* The degree day figures should not be taken from tables for this purpose as they will vary according to the heat loss and internal gains of the building. (Comparisons of several different sources of mean daily temperatures for each month are suggested.) 'Normals' have been calculated over different time spans (e.g. 10, 15, or 30 years) and because of changing weather patterns, and possibly cleaner atmospheric conditions due to more stringent pollution regulations, the more recent the information the better. This data set, for the worked example, is listed in Appendix E, Table E.12.

The second set of climatic data is required to assess whether or not the energy collection space is in danger of overheating on a clear winter's day. For this reason the month within the heating season with the highest solar gains and highest outdoor temperatures is chosen. It is suggested that either October or March should be the month modelled (see Fig. 3.7, Section 3.3). The data required are the mean hourly outdoor temperatures for a typical day in the month to be modelled. These temperatures are used in the overheating assessment methods outlined in Section 3.14 ('Overheating').

* See Section 3.11 ('Heating Load Calculations') for more information on degree day figures.

27

The values used for the worked example are shown in Appendix E, Table E.9. If these temperatures are not available but the mean daily minimum and maximum temperatures are, then the former can be approximated to a high enough degree of accuracy for the overheating design method. A sinusoidal graph is drawn using the mean daily minimum and maximum values for the month chosen as the minimum and maximum points on the graph respectively. The minimum daily temperature usually occurs before sunrise, the maximum in mid-afternoon. Equation (31) in Appendix B may be used to derive a regular sinusoidal temperature fluctuation.

3.2.   THE SITE AND ITS MICROCLIMATE

Vernacular architecture is highly responsive to its immediate environment because the owners and builders understood their environment better than we do ours today. This is due to several differences in lifestyle: today we move house more frequently and fewer of us work out of doors. Thus we do not understand from experience how much windier it can be in one part of a field than in another. Today we must do more investigation of the site using instruments to help us make such decisions.

The building site plays an important role in determining the energy requirements of a building. Both the summer ventilation or cooling loads and the winter heating load can be reduced by well-considered site use. Close observation is required to highlight the preference areas on a site. Existing vegetation, geology and topography all play a part in creating a unique microclimate for every site. There are analytical tools available for simulating the wind flow patterns around buildings, trees and landforms and for on-site investigations of the microclimate.

Observation of the site when it is under a layer of snow may reveal areas where snow has melted first. This will indicate a warm patch which could be employed for the building site.

Wind speed measurements should be taken at different points on the site in winter in an attempt to detect sheltered areas. Wind speeds are measured using a simple anemometer.*

A second useful instrument is a black body thermometer. This will measure the mean radiant temperature. Such measurements at different locations on a site will indicate warmer areas and in itself this provides a useful exercise in learning what factors affect the microclimate.

* Obtainable from Edmund Scientific Co., Edscorp Building, Barrington, New Jersey 08007, USA.

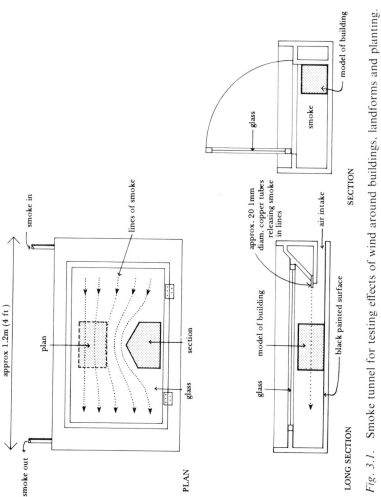

Fig. 3.1. Smoke tunnel for testing effects of wind around buildings, landforms and planting.

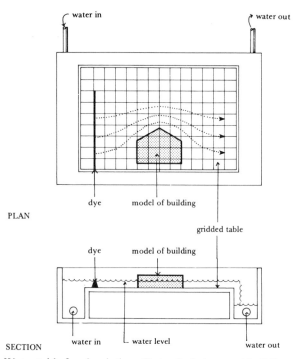

Fig. 3.2. Water table for simulating effects of wind around buildings, landforms and planting.

The use of trees and landforms as shelter belts is one of the most powerful aids available for influencing the microclimate. A wind tunnel is the obvious design tool for investigating the effects of shelter belts, but would be outside the budget of an architectural practice. It is possible to construct an inexpensive smoke tunnel or water table as shown in Figs 3.1 and 3.2. These will provide useful design information but the inherent inaccuracies of such tools should first be understood.

The Reynolds Number. The value of the Reynolds Number is in its use for comparisons between a real-life situation and testing methods. It is computed as follows:

$$\text{Reynolds Number} = \frac{v \times D}{\gamma} \tag{1}$$

where v = velocity, m/s (ft/s)
  D = dimension under consideration, m (ft)
  γ = viscosity, m²/s (ft²/s).

In real life the Reynolds Number will be above 2000, which is approximately the critical figure for laminar flow—above this figure there is a breakdown of layer boundaries and turbulence occurs. In the graphic testing methods suggested below the Reynolds Number will be far less than 2000, otherwise it would be impossible to distinguish the smoke or dye paths.

(i)   Real-life situation
       Assume a wind velocity of 10 mile/h $\simeq$ 5 m/s (15 ft/s)
       Assume the dimension considered is 7 m (21 ft)
       γ for air is 0·014 m^2/s (0·14 ft^2/s)
       Therefore Reynolds Number = 2500 (2250)

(ii)   Smoke tunnel
       Velocity = 1 m in 4·5 s (2 ft in 3 s) $\simeq$ 0·22 m/s (0·6 ft/s)
       Dimension = 0·13 m (0·4 ft)
       γ = 0·014 m^2/s (0·14 ft^2/s)
       Therefore Reynolds Number = 2·0 (1·7)

(iii) Water table
       Velocity = 1 m in 5 s (3 ft in 5 s) $\simeq$ 0·2 m/s (0·5 ft/s)
       Dimension = 0·15 m (0·5 ft)
       γ = 0·19 m^2/s (1·9 ft^2/s)
       Therefore Reynolds Number = 0·16 (0·13)

This demonstrates the order of magnitude difference between a real-life situation and these test methods.

A second area of inaccuracy in both the smoke tunnel and water table methods is that they do not imitate the behaviour of wind passing over land, where there is an increase in velocity with height above the ground. This effect is duplicated in a wind tunnel but is difficult to achieve in the simple equipment presented here.

A third area of inaccuracy is as a result of being able to model forms in either plan or section only. The effect of wind around a three-dimensional form would be rather different from that seen around a sectional form.

Thus the usefulness of these methods lies only in the graphic representation of the wind effects around such forms. By seeing the behaviour of the wind due to changing model forms it is possible to visualize the principles involved. Although there is inherent inaccuracy, with increased accuracy it would be impossible to see these effects graphically.

When experimenting with these methods only one change (in the form)

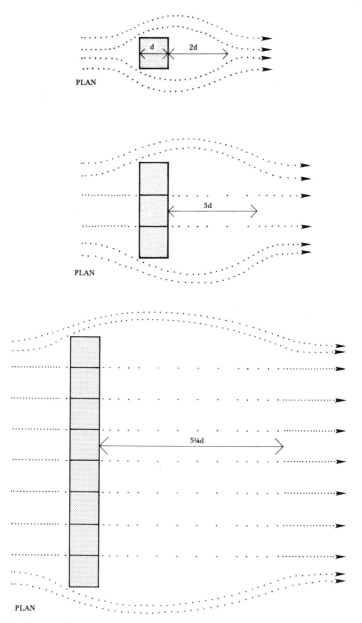

Fig. 3.3. Increasing the depth of shelter is possible by means of increasing the length of the shelter belt (plan).

should be performed each time so that a cause-and-effect sequence can be built up.

The smoke tunnel was found to give more realistic wind effects than did the water table. An additional use for both methods is their ability to study summer ventilation within the house form and to demonstrate the effects of changing size of openings and placement of internal partitions.

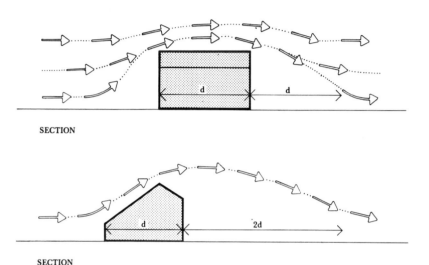

SECTION

SECTION

Fig. 3.4. The sheltered area increases in height and depth when the shelter belt is shaped to lift the wind.

Certain wind shelter patterns can be reduced to rules-of-thumb and these are indicated in Figs 3.3 and 3.4.

The type of tree used to provide a shelter belt should be dense and coniferous so that a barrier is provided in winter. Smaller trees and shrubs planted immediately on the windward side of the shelter belt will assist in lifting the wind up and thus increase the area of shelter. Observation of trees in similar locations will help to select those which will flourish in windy locations and in similar soil conditions.

Other important factors of on-site investigation are dealt with in Sections 3.4 and 3.12 ('Solar Interference Boundaries' and 'Fabric Losses'). These relate to forms which will obstruct the view of the sun and to the height of the groundwater table. Further information on site investigation will be found in references [1] and [2].

3.3.   ORIENTATION AND TILT OF THE COLLECTING SURFACE

The limitations of orientation are not as stringent as they may first appear. The design solution will emerge from many such technical limitations and perfection within any is not the goal, otherwise design becomes slave to technology. The work of Balcomb and others [3] shows that although, as

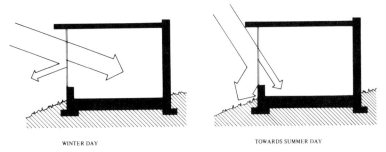

WINTER DAY   TOWARDS SUMMER DAY

Fig. 3.5.   Diagrams of the effect of Fresnel's Law.

expected, a due south orientation is optimum, variations of 30° east or west will reduce performance by only about 3%. Cloud-cover patterns may change these conclusions; a locality may have more overcast afternoons than mornings, thus favouring a slightly east of due south orientation.

The optimum tilt for passive solar heating differs from that for an active system. A vertically glazed collecting surface is preferred in higher latitudes. The low altitude of the sun in winter is more normal to the glass which, following Fresnel's Law, enables penetration of a greater proportion of the incident direct solar energy in the months when most heating is required (Fig. 3.5).

As shown in Fig. 3.6, vertical glazing presents a larger aperture to the sun during the winter months, and a smaller one in the summer months when the energy must be rejected. Thus vertical glazing is almost a self-regulating device for passive solar heating (Fig. 3.7). A reflecting surface placed on the ground in front of vertical glazing and of equal area to it will, with every low cost, increase the collection area by approximately 40%.

Horizontal collecting surfaces are used in lower latitudes where the winter solar radiation arrives at a vertical plane with an angle of incidence greater than 50°. These high angles of incidence greatly reduce the quantity of solar radiation transmitted through vertical glazing (Fig. 3.6).

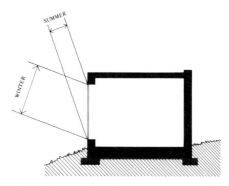

Fig. 3.6.   Varying aperture between winter and summer.

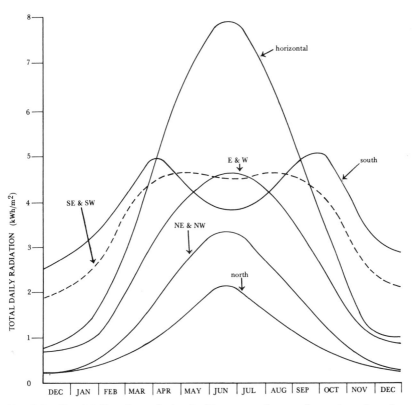

Fig. 3.7.   Total daily solar radiation incident on various surface orientations for a clear day in London (Kew); ground reflection included [4].

Horizontal collecting surfaces will require a device to prevent solar gains during summer.

Glazing with less than a 90° tilt must be incorporated into a design with great caution. It will present major overheating and shading problems in summer, and high convection and radiation heat losses in winter by allowing radiative contact between the thermal storage area, glazing and a large area of the sky vault.

Glazing with greater than a 90° tilt may be an advantage in latitudes greater than 50° to reduce the need for extensive summer shading devices, but may be counter-effective in overcast climates where so reduced a view of the sky vault may also reduce heat gains.

3.4.   SOLAR INTERFERENCE BOUNDARIES

The altitude and azimuth angles of the sun change as it apparently moves across the sky. The Sun Angle Calculator® is an indispensable tool for predicting these and the profile angles of the sun throughout the year for a variety of latitudes.* To determine the contribution of solar energy towards the heating requirements of a building it is necessary to know whether the sun will be obscured at any time of the day by buildings, trees or landforms. If the sun is obscured then diffuse solar energy will be the only contribution towards the heating requirements.

The ability to see the sun at all times during the day is another requirement of solar heating to be looked upon as important but not of overriding importance. The optimum solution to this may not be the best design solution. It is when aesthetics, function and economic feasibility are weighed against each other that the design solution emerges.

In all there are four methods presented here for determining solar interference boundaries, and each has its own merit.

(a) Figure 3.8 is a graph showing solar interference for 21 February and 21 October for a latitude of 52 °N. The graph is constructed to show where various vertical heights will interfere with the view of the sun from the point of origin. The contours are for the vertical distances marked, above ground level. The point of origin of the lines of azimuth for each hour is the point being modelled, assumed to be 2 m (6 ft) above ground level at the surface of the building. The graph method [5] is used by laying a tracing of the

® Available from Libbey–Owens–Ford Company, Toledo, Ohio, USA.
* 'Profile angle' is explained in Section 3.9 ('The Solar Modulator').

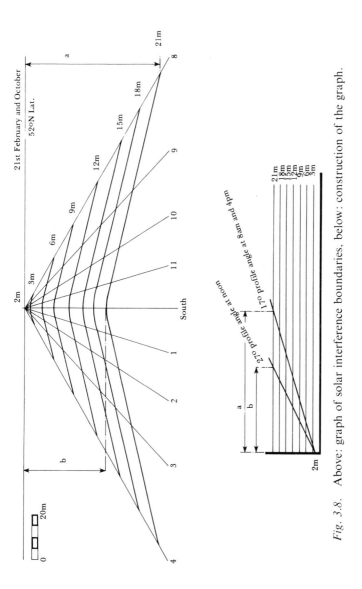

Fig. 3.8. Above: graph of solar interference boundaries, below: construction of the graph.

building and site plan (both drawn to the same scale) over the graph so that
the noon azimuth line is due south with respect to the orientation of the
building and site. The graph is constructed (Fig. 3.8) by drawing a section of
the building face with the profile angles for sunrise and noon originating
from the point being modelled. The horizontal distance from the face of the
building to the profile angle lines drawn is measured and used in plan to
mark off the interference boundaries along the respective lines of azimuth.
This distance in plan is measured at right angles to the east–west axis. (For
an application example, see Appendix E, Fig. E.1.)

Fig. 3.9.   The Heliodon and scaled model simulating 21 December at 9.00 a.m. at
52 °N latitude.

(b) The sundial-type Heliodon and scaled model can simulate the
apparent movement of the sun through the day for each month (Fig. 3.9).
The Heliodon can be adjusted to model any latitude. It may be constructed
from a photocopy of Fig. 3.10 and mounted on an east–west axis with
respect to the orientation of the building model. Either a parallel-ray
artificial light source or the sun must be used. This method will be more
useful than the graph method if the building has a contorted plan shape.
For accurate results the mounting must be on an adjustable surface,
capable of being clamped. To facilitate estimation of area of window in sun,
at any hour, it is suggested that the windows be divided into ten equal
rectangles. In this way a percent recording can be made of each window for
every hour modelled. The readings should be taken on the hour to average
the percent area in sunshine over the hour centred on the hour (i.e. 9.30 to
10.30 is averaged by 10.00). This is because hourly insolation values listed
are usually treated in this manner.

(c) Computer simulation of the interaction between building shape and solar position for each hour of the day gives good results and is capable of providing great flexibility in building design and window size and placement, without the work involved in making scaled models of the building. This method was adopted in the worked example presented in Chapter 4. It is based on trigonometry and is commonly used for commercially offered computer simulation packages for assessing the energy requirements of buildings. Thus it offers nothing new to the field and consequently will not be presented here. An example of the data gained is shown in Appendix E, Table E.14.

(d) If the site can be visited there are commercially available sunlight-finding instruments. These will tell in minutes which objects will obscure the sun, and when. One such instrument has been developed by J. C. C. Williams* and is illustrated in Fig. 3.11. The advantage of such an instrument is that it may be adjusted for use at any latitude.

Glue a photocopy of the following sheets onto light cardboard (e.g. a manilla folder) and cut out the parts.

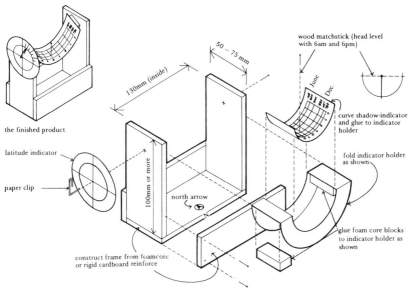

Fig. 3.10. A sundial-type Heliodon—construction sheet.

* Details of availability from J. C. C. Williams, 4 Frownen Terrace, Cradoc Road, Brecon, Powys LD3 9HB, Wales.

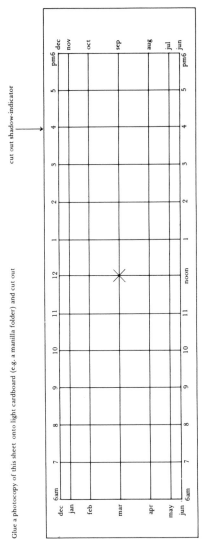

Fig. 3.10.—contd.

Glue a photocopy of this sheet onto light cardboard and cut out parts

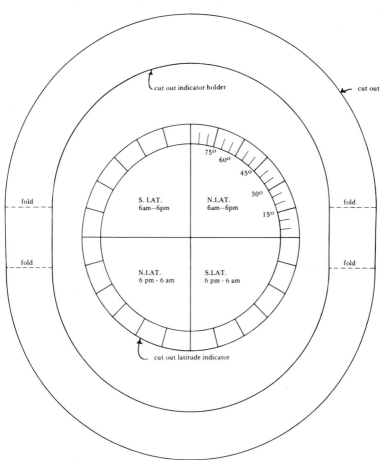

Fig. 3.10.—contd.

Fig. 3.11. Sunpath-finder by J. C. C. Williams.

3.5. SOLAR DATA

Again two sets of data are required.

(a) Having determined when the sun is being obscured, it is necessary to calculate the resulting monthly total solar gains through the windows and walls for each month of the year. Only with the help of a computer can this be done on a day-by-day basis. Thus the 21st day of each month is assumed to be a typical day for that month. Tables are available [6, 7] giving the hourly radiation on various surface orientations for a clear day at a given latitude. Care should be taken to determine whether or not the values listed are for insolation on a surface or for insolation received through glazing.

If a window is in shade, only diffuse radiation will be counted; the data may be taken directly from the west-oriented vertical surface values (before noon), and the east-oriented vertical surface values (after noon), in the tables for clear day insolation. No correction is necessary because an overcast sky will, if anything, contribute more radiation than a clear north sky; summer mornings and evenings are an exception where, in high latitudes, the sun enters the northern hemisphere.

If a window is potentially seeing the sun, the clear day insolation tables will not allow for the days when the sky is partly clouded. Consequently a correction must be made to these figures. The rule-of-thumb correction method presented by Anderson and Riordan [8] was found to be inaccurate when applied to the English data in the worked example (Chapter 4). The correction factor can be more accurately calculated if one has access to values of the mean monthly or mean daily total solar radiation. The correction factor for any one vertical orientation was found to be different from another and different again from the horizontal surface correction factor. Also, these correction factors must be determined for each month. Considering a surface with a particular orientation and tilt, eqn (2) will give the correction factor for that month.

$$Cl = \frac{I_{\text{tav}}}{I_{\text{tcl}}} \tag{2}$$

where Cl = the correction factor for that month,

I_{tav} = the mean monthly total insolation on the surface considered,

I_{tcl} = the monthly clear day total on the surface considered [6, 7].

If one has figures for the mean daily total solar radiation for each month, the method is identical except that the daily values replace the monthly values in eqn (2).

It may be the case that the only total radiation figures given are for a horizontal surface, listed as hourly total and diffuse radiation values, and that the building is oriented in a direction other than due south. Converting these horizontal data into insolation values on any vertical orientation is possible, although cumbersome unless a programmable calculator is used.

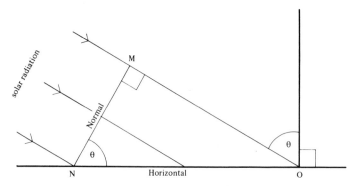

Fig. 3.12. Illustrating the Cosine Law.

The method is as follows:

$$\text{Total Solar Radiation} = \text{Direct} + \text{Diffuse Solar Radiation} \qquad (3)$$

Considering first the direct portion of the total solar radiation, the Cosine Law (Fig. 3.12) states:

$$I_{dh} \times NO = I_{dn} \times NM \qquad (4)$$

where I_{dh} = the direct solar radiation for that hour on the horizontal surface NO,

I_{dn} = the direct solar radiation for that hour on the surface NM, normal to the solar radiation.

Therefore:

$$I_{dn} = \frac{I_{dh}}{\cos \theta} \qquad (5)$$

where θ = the angle of incidence (which for a horizontal plane = $90° -$ altitude).

In the same way, by simple trigonometric relationship:

$$I_{dv} = I_{dn} \times \cos \psi \times \cos W \qquad (6)$$

where I_{dv} = the direct solar radiation for that hour on the vertical surface,
ψ = the angle of altitude,
W = the angle between the sun orientation and the normal to the vertical surface, known as the sun–wall azimuth; varies every hour.

In the previous equations only unit areas are considered in the relationships between I_{dh}, I_{dn} and I_{dv}.

The diffuse portion of the total solar radiation must now be considered. A horizontal surface sees the total sky vault; a vertical surface sees only half the total sky vault. Thus a rough approximation of the diffuse energy falling on a vertical surface is half that falling on a horizontal surface.

From eqn (3):

$$I_{tv} = I_{dv} + (0.5 \times I_{fh}) \tag{7}$$

where I_{tv} = the hourly total solar radiation falling on a unit area of vertical surface,
I_{fh} = the diffuse solar radiation for that hour falling on a unit area of horizontal surface.

Incorporating eqns (5) and (6) into eqn (7):

$$I_{tv} = ((I_{dh}/\cos\theta) \times \cos\psi \times \cos W) + (0.5 \times I_{fh}) \tag{8}$$

The results of eqn (8) for every hour the sun is up on the 21st day of the month are summed to obtain the mean daily total solar radiation on the vertical surface of that orientation. Then the correction factor can be computed from eqn (2). The worked example shows this information in Appendix E, Table E.16.

The monthly solar gain totals, through windows and walls, are computed by the methods outlined in Section 3.8 ('Calculating Solar Gains').

(b) The second set of solar data required is that for predicting the possibility of overheating. The hourly clear day intensities of total solar radiation per unit area of the vertical surface orientation and month being modelled are taken from the tables [6, 7]. Allowances are made for the results obtained from the design methods in Section 3.4 ('Solar Interference Boundaries'). See Appendix E, Tables E.2 and E.7, for these values in the worked example.

Further information on solar gains for the overheating design method is presented in Sections 3.8 and 3.14 ('Calculating Solar Gains' and 'Overheating').

3.6. WINDOWS: IMPROVEMENT FOR SOLAR COLLECTION

North windows and horizontal roof lights have high heat losses and the size and number of these should be minimized.

This section deals primarily with the large areas of south-facing glazing which are necessary for solar energy collection. Double glazing will reduce heat losses but to increase thermal resistance even more becomes a substantial problem. This problem can be tackled in two ways: by means of movable insulation or by means of improving the thermal performance of the transparent membrane. The number of economically attractive and well-performing solutions to this problem are few.

Movable insulation comes in various forms:

(a) Curtains or Drapes

Depending on the materials used and the treatment of the edges of the curtains these can be useful. New insulating materials, low emissivity blinds and multiple layer foil blinds are coming on to the market.

(b) Nightwall

Rigid expanded polystyrene sheets inserted within the window surround and held in place by magnetic strips. These are light, easy to apply and inexpensive but require storage space unless built in the form of shutters or concertina blinds. Edge treatment is again critical to their performance.

(c) Mechanical Methods

Under this heading come Beadwall,* Skylid,* and external movable insulation panels.

Beadwall [9] is a system whereby polystyrene beads are pumped into the cavity between double glazing at night or during cloudy spells, and removed when the sun is shining. The pump exerts considerable pressure on the glazing during this process. Both single and double strength glass failed to withstand this pressure. Float glass of 5 mm ($\frac{3}{16}$ in) and plate glass of 6 mm ($\frac{1}{4}$ in) thickness have been used without problems. However, should the system become clogged and internal pressure builds up, these glasses will explode. Therefore, to avoid such potential hazards, tempered safety glass or fibreglass is recommended. Fibreglass is the obvious choice on the grounds of cost but it degrades with age and an ultraviolet sacrificial

* Patented systems by Zomeworks Corp., PO Box 712, Albuquerque, New Mexico 87103, USA.

coating must be applied and maintained. Also fibreglass is not transparent and thus conflicts with any view requirements. The final point about this system is that it is very expensive: in excess of £50/m² ($8/ft²). The U–value of the system when full of beads = 0·6 W/m² °C (0·1 Btu/h ft² °F); when empty = 2·8 W/m² °C (0·5 Btu/h ft² °F).

Skylids can be used on both skylights and vertical glazing. The 300 mm (1 ft) wide insulated louvres are automatically opened or closed by a sun-sensitive system consisting of coupled freon-containing cans. One can is on either side of the louvre. When the sun emerges the energy absorbed by the first can will expand the freon and move it into the second can. The resulting imbalance in weight between the cans will cause the skylids to open. The seals around the edges of the skylids are not airtight and this reduces its potential thermal insulating performance. This is another expensive system: £82/m² ($13/ft²). The U–value when the skylids open (dependent on position) = 2·8 W/m² °C (0·5 Btu/h ft² °F); when closed = 1·14 W/m² °C (0·2 Btu/h ft² °F).

External movable insulation panels must be robust enough to withstand the elements. Good seals are again difficult to achieve. If the panels are of the type that lower away from the building, they can perform the additional function of a reflector, increasing the incident solar radiation on the glass. The cost of these panels is high but cost and U–value will be dependent on their construction. The U–value attainable at night = 1·25–0·85 W/m² °C (0·22–0·15 Btu/h ft² °F).

In all the systems described above, with the exception of Beadwall, there is the problem of attaining airtight seals. Without good seals on all four edges the air against the glass surface will cool and fall into the room, reducing the potential thermal performance of the system. Depending on price and insulating capabilities, shutters may well prove to be one of the more competitive means of reducing heat losses through windows, especially in high latitudes where the night-to-day ratio in winter is about 3:1. But the time commitment needed on the part of the user for opening and closing most of the systems described above is considerable. Furthermore the one basic disadvantage of all movable insulation systems is that there is no increased resistance to heat losses during solar collection periods. This and other design requirements of windows are being met by materials now being developed in many countries throughout the world.

To appreciate the methods of improving the thermal performance of windows by the use of glazing alternatives it is necessary to understand how the heat is being lost. The original 'greenhouse effect' concept is that

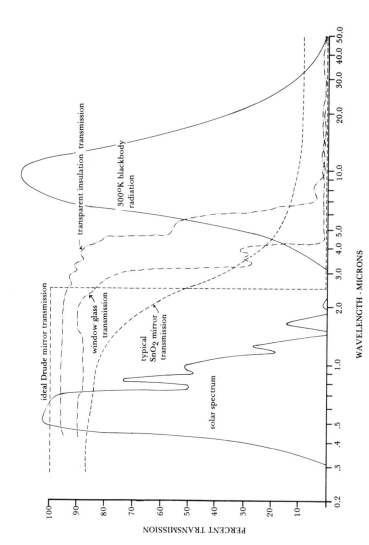

Fig. 3.13. Transmission values of various membranes and superimposed radiation curves (10).

because glass is transparent to the solar radiation wavelengths and opaque to infrared (longwave) radiation it acts as an energy valve. Figure 3.13 shows, among other things, the different wavelengths of solar and room temperature radiations. However, it turns out that, although glass does have these properties, it is little more than a convection trap. For when polythene, which is fairly transparent to the infrared region, is tested against glass there is no discernible difference between the temperatures produced by glass and those by polythene.

The routes of energy escape are different: the energy leak in glass is in its absorption of infrared radiation and then reradiation of this energy to the sky, whereas with polythene the majority of the heat lost is through its transmission window at the infrared wavelengths. With either type of transparent cover, the quantity of energy escaping is 60% of that escaping from an equal area of uncovered ground.

For any material the following two equations are true within a given wavelength:

$$\alpha + R + \tau = 1 \qquad (9)$$

and

$$\alpha = \varepsilon \qquad (10)$$

where α = % absorption, R = % reflection, τ = % transmission, and ε = % emission, and where percentages are expressed as fractions.

Thus it follows that, because glass is opaque to infrared radiation and because its reflection of infrared radiation is negligible, it absorbs almost all and then emits almost all of the infrared radiation absorbed.

The overall U–value (Table 3.1) of single glazing is not greater than 7.38 W/m^2°C (1.3 Btu/h ft^2°F) and when storm windows are added the U–value is approximately halved to 3.18 W/m^2°C (0.56 Btu/h ft^2°F). Approximately 60% of the heat loss from the double-glazed window is in the form of infrared radiation.

There are many properties to be considered, other than thermal performance, when comparing alternative glazing materials; a few of these are transmission to solar radiation, durability, and distortion of the image received through the material.

There are several possible ways, other than movable insulation, of attempting to increase the overall solar collecting performance of the transparent membrane used:

(a)   treatment of the space between double glazing in an attempt to eliminate or reduce convective losses;

Table 3.1.   U–*values of windows* (*after* [11])

Orientation and exposure	Glazing configuration	U–value	
		W/m²°C	Btu/h ft²°F
S sheltered	Single glazing	3·97	0·70
	Double, 6·4 mm ($\frac{1}{4}$ in) space	2·67	0·47
	19·1 mm ($\frac{3}{4}$ in) or more space	2·32	0·41
S normal; W, SW, SE sheltered	Single glazing	4·48	0·79
	Double 6·4 mm ($\frac{1}{4}$ in) space	2·90	0·51
	19·1 mm ($\frac{3}{4}$ in) or more space	2·50	0·44
S severe; W, SW, SE normal; NW, N, NE, E sheltered	Single glazing	5·00	0·88
	Double 6·4 mm ($\frac{1}{4}$ in) space	3·06	0·54
	19·1 mm ($\frac{3}{4}$ in) or more space	2·67	0'47
W, SW, SE severe; NW, N, NE, E normal	Single glazing	5·67	1·00
	Double, 6·4 mm ($\frac{1}{4}$ in) space	3·29	0·58
	19·1 mm ($\frac{3}{4}$ in) or more space	2·84	0·50
NW severe	Single glazing	6·47	1·14
	Double, 6·4 mm ($\frac{1}{4}$ in) space	3·58	0·63
	19·1 mm ($\frac{3}{4}$ in) or more space	3·00	0·53
N severe	Single glazing	7·38	1·30
	Double, 6·4 mm ($\frac{1}{4}$ in) space	3·80	0·67
	19·1 mm ($\frac{3}{4}$ in) or more space	3·18	0·56

(b)   increasing the transmission to solar radiation;

(c)   lowering the longwave (infrared) radiation losses by reflecting them back into the room.

These three methods will be discussed in greater depth below.

(a) By tackling convective losses alone, the potential energy savings in reducing radiational losses are missed. Only 40 % of the heat lost through double glazing is by convection. Within this category there are two approaches taken: eliminating convection and conduction by evacuating the air space, or inhibiting heat flow by introducing a suitable gas in the interpane space.

Even a partial vacuum, to the extent of reducing convection, would create sufficient pressure (4 lb/in²) to warrant a closely spaced structural system within the air space [10]. This structural system would be expensive, would distort the image received, increase losses due to conduction and increase absorption of solar radiation. Absorption increases, and thus transmission decreases (eqn (9)), by approximately 1% for every 1 mm increase in thickness of normal glass.

The introduction of a heavy gas to reduce thermal conduction and convection within the interpane space has been attempted. However, it is uncertain in the literature which is being dealt with, because low convection and low conduction call for opposite molecular weights [10].

(b) The transmission of glass to solar radiation will not appreciably deteriorate with age, whereas many plastics exhibit proliferation of colour centres on exposure to sunlight in a matter of months. A single pane of 4 mm ($\frac{1}{6}$ in) low-iron glass will have a transmission of 92%* to solar radiation compared with 87% for normal glass. Absorption of solar radiation is increased and transmission decreased if iron is present in the glass. Iron causes a green tint to the glass when viewed from the edge of the pane. Double glazing with low-iron glass will give a transmission of 85% (92% × 92%) compared with 75% (87% × 87%) for double-glazed normal glass. The additional cost of low-iron glass over normal glass is approximately 25%.

The property which determines the quantity of solar radiation reflected is the index of refraction. This is a surface effect that has a bearing on the quantity of solar radiation transmitted through the material (eqn (9)). The index of refraction for glass is 1·5: anything lower than this will have a higher transmission to solar radiation and will accept a larger angle of incidence, with greater efficiency.

Teflon FEP® is a flexible film produced by Du Pont. It comes by the roll, is 2 mil thick and costs £3·15/m² (50 c/ft²). Teflon FEP® is inert and has an index of refraction of 1·33 which gives it a 96% transmission value for solar radiation [10]. Thus one could have six layers of Teflon FEP® and still get a slightly higher transmission value for solar radiation than double glazing. This type of performance is not possible using normal glass because the resulting increase in thermal resistance is less than the decrease in transmission. The use of Teflon FEP® in effect reduces convection losses in a similar way to the concept involved in a honeycomb lattice.

These are the disadvantages to Teflon FEP®: it is not rigid and must be protected both internally and externally, and a 2 mil film of it has 40% transmission to longwave (infrared) radiation. The result is that the gain from decreased convective losses is cancelled by the increased radiation

---

* Unless otherwise specified, transmission values are for solar radiation normal to the surface.

losses. Thus Teflon FEP® used in this way is only of value because of its slightly lower cost than glass because it is cheaper to install. Installation is effected by fixing all four edges and heat-shrinking it using a blow heater. There is slight distortion of the image seen through it.

Teflon FEP® can be used as an anti-reflection coating to glass. Glass is dipped into an aqueous slurry of Teflon FEP® and put into a fusing oven at 370°C (700°F). Ordinary glass coated on one side will have its transmission to solar radiation increased from 87% to 92%. The coating can be marked too easily for external use, thus glass will be coated on one side only. This process has not yet been tried, and costs, although expected to be minimal, are not known. The use of this coating on low-iron glass would improve its performance also.

Suntek Inc., California, is developing a transparent insulation by coupling Teflon FEP® to a layer of transparent material which is opaque to infrared radiation. The transmission to solar radiation is reduced by only 1% from the value of Teflon FEP® to 95% (Fig. 3.13). By using three layers of this between low-iron double glazing (all separated by 2-in air spaces), it is possible to reduce convection losses and obtain an overall U–value of 1·25 W/m² °C (0·22 Btu/h ft² °F) and transmission to solar radiation of 73% at a cost of £9.45/m² ($1.50/ft²) above the cost of double glazing (excluding framing). The disadvantages are that there is some visual distortion to the image seen through this transparent insulation and the frames become large in section and thus cumbersome and expensive to produce.

Multiple layer transparent insulation goes somewhat further than reducing convective losses; it also reduces losses due to infrared radiation by increasing the number of absorption and emission steps to the outside.

(c) The third method of attempting to increase the overall solar collecting performance is the Drude mirror or heat mirror. The ideal transmission curve of a heat mirror is shown in Fig. 3.13. Heat mirrors have been on the market for 20 years but at a very high price. Typical applications have been for aeroplane and train windscreens. There are two approaches to producing heat mirrors. The first is by etching very fine lines on the glass, so fine that they will reflect the longwave (infrared) radiation. The second uses the same concept but, instead of using lines, thin inorganic layers are applied to the transparent substrate. With both methods there is little reduction in the transmission of solar radiation. The costs are high because both are batch methods. Research is being carried out in several

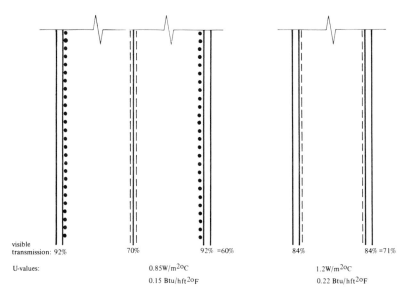

Fig. 3.14. Two possible assemblies of the heat mirror being produced by Suntek Inc., California.

laboratories throughout the world to produce a heat mirror by an on-line method. This has been achieved by Suntek Inc. The substrate being used is mylar. Only 300 mm (1 ft) widths are available at the moment and the price is fairly high. These strips can be bonded together or applied to a glass surface. The two possible assemblies and their properties are shown in Fig. 3.14.

The expected additional cost due to the heat mirrors in both assemblies is the same: £3·00/m² (48c/ft²) (fixing included). Heat mirror samples show that there is no visual distortion to the image received through it. But they do show signs of surface corrosion. This defect is expected to be overcome within a year.

Flachglass AG in Frankfurt, Germany, produce a heat mirror or low-emissivity glass (eqn (10)) by means of a gold deposit on the glass.* The system manufactured by them consists of 6 mm glass, 12 mm gap filled with a heavy gas, and 6 mm glass. The overall U–value is 1·6 W/m² °C

* Distributed in England by Bomert, Teves and Blankley Ltd, Pembroke House, 44 Wellesley Road, Croydon CR9 3PD.

(0.28 Btu/h ft^2 °F), solar transmission is 60% and the system is sold pre-sealed. The cost of this product varies depending on the size of glazing—for 1 m^2 the cost is £40·00 ($8/ft^2).

The on-line production of a non-corroding heat mirror has been successfully developed by Airco, a US company owned by British Oxygen Company (BOC International). Airco and Pilkington Brothers Ltd will shortly be setting up a plant in Sweden to produce this glass. It will have a blue tint rather than the gold tint of the Flachglass product and it promises to be considerably less expensive. The coating is scratch resistant and non-corrosive and therefore there is no need to have hermetically sealed units.

All the methods of improving the solar collection performance of windows discussed in this section, can be assessed in terms of cost by the program presented in section 3.18 ('Cost Analysis').

3.7. THERMAL MASS

Rules of Thumb

(a) *Non-diffusing Direct Gain. Primary thermal mass: for concrete the maximum useful thickness when energy is applied to one side only is 225 mm (9 in). Thermal capacity should be sized as close to 240 Wh/m^2 °C (45 Btu/ft^2 °F) of the south-facing collection window.*

Primary thermal mass: area of phase change material tile (Sol–Ar–Tile®) should be 1·75 times the area of the south-facing collection window.

Secondary thermal mass: the maximum thickness to be employed in daily thermal exchange is 150 mm (6 in) of concrete when the area of secondary thermal mass is 0·25 times the area of primary thermal mass. Similarly:

> *100 mm (4 in) for equal areas*
> *50 mm (2 in) for five times the primary thermal mass area*

(b) *Diffusing Direct Gain. Thermal mass should be sized at 290 Wh/m^2 °C (55 Btu/ft^2 °F) of floor area* [12].

(c) *Trombe–Michel Wall. Balcomb's recommendations are that Trombe–Michel walls should have a minimum thermal capacity of 160 Wh/m^2 °C (30 Btu/ft^2 °F) of south-facing glazing. This is equivalent to 300 mm (1 ft) thick concrete walls or 150 mm (6 in) thick water walls.*

Time lag: on a clear day for a solid concrete wall with double glazing on the outside [13]:

Thickness	Inside surface swing	Time lag (h)
200 mm (8 in)	22 °C (40 °F)	6·8
300 mm (12 in)	11 °C (20 °F)	9·3
400 mm (16 in)	5 °C (9 °F)	11·9
500 mm (20 in)	3 °C (5 °F)	14·5

(*d*) *Active or Hybrid System. Thermal mass in contact with the collection space by means of a convective loop can be sized in a similar way to that used for an active system:* 80 Wh/m² °C (15 Btu/ft² ° F) *of south-facing glazing. It should be remembered that only a thin layer of the convective thermal mass will partake in the daily storage of heat (approx. 25 mm (1 in)), unless water is used.*

Thermal mass is the means of heat storage and air temperature stabilization that passive design employs. Most of the energy received from the sun is in the form of shortwave (visible) radiation and thus it is the surface colour of the area it strikes that controls the quantity of energy absorbed or reflected. Of the energy absorbed by the thermal mass, some will be conducted into the thermal mass and the remainder will be lost from the surface by convection and radiation. Convective losses will directly increase room air temperatures but radiated energy cannot be transformed to heat until it strikes an object or surrounding surface. If the objects it strikes are furnishings and the surrounding surfaces have low thermal capacities (specific heat × density) or low conductances (reciprocal of the resistance), their surface temperatures will rise quickly, causing large heat losses to the room air. In such a situation overheating will soon occur. There are two methods of handling the incident solar energy as it enters the collection space: by diffusing the light and thus scattering it to all areas of thermal mass throughout the space; and by absorbing the majority of the solar radiation at the area in direct contact with the non-diffused solar radiation.

(*a*) *Diffusing Direct Gain System*
This involves a conscious effort to diffuse the sunlight so that it is scattered to all areas of thermal mass. It is achieved either by using diffusing glass or by allowing the solar radiation to reflect diffusely off a white painted masonry wall. This method is exemplified in the St George's Secondary

School extension at Wallasey in England. It is a method of great significance since it effectively reduces the surface temperature of the thermal storage mass and by so doing reduces the air temperature fluctuations within the room. It is both interesting and instructive that E. A. Morgan, in his patent application document [12], stated that a thermal capacity of 55 Btu/ft^2 °F of floor area was needed for optimum performance of his system. This is a good rule-of-thumb for diffusing direct gain systems if one remembers that any thermal mass in excess of 225 mm (9 in) will not take part on a day-to-day basis.

The potential constraints of this type of system are the obscured view through the south glazing or the need to step the roof in order to effect the diffuse reflection. Matt-white louvres could be used to attain both diffused solar energy and unobscured view. In climates with prolonged and high levels of solar radiation it is suggested that this type of system be used with caution. Control of the air temperature fluctuations in such areas will be possible only by means of dark interiors rather than the white interiors used at the Wallasey School.

(b) Non-diffusing Direct Gain System

In this system there is a primary thermal mass which is the target area hit by direct solar radiation. This area is designed to absorb most of the incident solar radiation by having a dark surface colour. The main method of preventing overheating in this type of system is by the use of longwave radiative transfer of energy to secondary thermal mass. The surfaces surrounding the target area are constructed with materials which have high thermal capacities and high conductances. Thus the energy-primed secondary thermal mass stores heat and consequently tames the fluctuations in room air temperature. Unlike the absorption requirements of visible wavelengths, absorption of the energy radiated from room temperature surfaces is not influenced by colour. The molecular structure of the absorber surface does, however, control the quantity of infrared radiation absorbed. All pecularities of radiation absorption are governed by the relationships described in eqns (9) and (10). Thus, because polished metal has a low emissivity of infrared radiation it would be a poor choice for the surface of secondary thermal mass (eqn (10)). Fortunately in this case, most building materials are good absorbers of infrared radiation.

Secondary thermal mass in indirect view of the target area can also be used to good effect for energy storage (Fig. 3.15).

In non-diffusing direct heat transfer passive systems an emphasis is placed on infrared radiation exchange, because after the absorption of

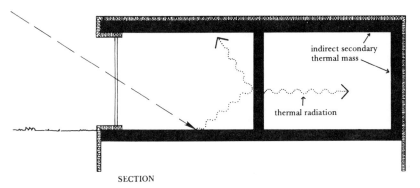

Fig. 3.15. Positioning of indirect secondary thermal mass.

direct solar energy this is where most of the action occurs. The proportional heat loss paths from room surfaces shown in Table 3.2 illustrate this point.

Computer simulations of the performance of non-diffusing passively solar-heated buildings were conducted using 'Thermal', the program developed by T. E. Johnson at MIT. Table 3.3 is an example of the data used for modelling. It will be noticed that the outdoor climate is particularly severe and that concrete is used for both target area and secondary thermal mass; also that the program does not account for internal gains which, although they have an influence on room air temperatures, will not affect the conclusions drawn from the simulations presented here.

The program models a ceiling target situation with an 85 % absorption of the direct solar energy gains. The secondary thermal mass has the same resistance to heat loss as the target, in the program, and is assumed to be opposite and parallel to the target area (i.e. on the floor). Simulation of the movement of heat throughout the room is accomplished by dividing the materials involved into many separate layers and by iteratively analysing the heat flow between these layers. This form of program is known as a finite difference thermal network program.

Table 3.2. *Proportions of heat loss
paths from various surface tilts*

Surface	Convection	Radiation
Floor	46 %	54 %
Wall	40 %	60 %
Ceiling	30 %	70 %

Table 3.3. *An example of the data modelled in simulation studies using 'Thermal'*

		Hours of the day						
		1	2	3	4	5	6	7
Clear day vert. comp. of solar gains*	Wh/m² (Btu/ft²)	141 (45)	279 (89)	354 (113)	392 (125)	354 (113)	279 (89)	141 (45)
Percentage noon target in sun*		100	100	100	100	100	100	100
Outdoor temperature	°C (°F)	−6·1 (21)	−5·0 (23)	−4·4 (24)	−3·9 (25)	−3·3 (26)	−1·1 (30)	1·1 (34)

Morning start temp. 18·3°C (65°F); air change/h 5 522 m³ (195 000 ft³); no auxiliary heating

	Area	Conductivity	Density	Specific Heat**	Thickness
Target (area = noon target area in sun)	508 m² (5 440 ft²)	0·13 W/m²°C/m (0·76 Btu/ft²°F h/ft)	2 243 kg/m³ (140 lb/ft³)	0·23 Wh/kg/°C (0·2 Btu/lb/°F)	15·0 cm (0·5 ft)
Secondary thermal mass	508 m² (5 440 ft²)	0·13 W/m²°C/m (0·76 Btu/ft²°F h/ft)	2 243 kg/m³ (140 lb/ft³)	0·23 Wh/kg/°C (0·2 Btu/lb/°F)	9·0 cm (0·3 ft)

Area of weather skin 3 520 m² (37 668 ft²), U–value 0·45 W/m²°C (0·08 Btu/h ft²°F)

* See Section 3.8 ('Calculating Solar Gains').
** Conversion of kJ to Wh is obtained by dividing kJ by 3·6.

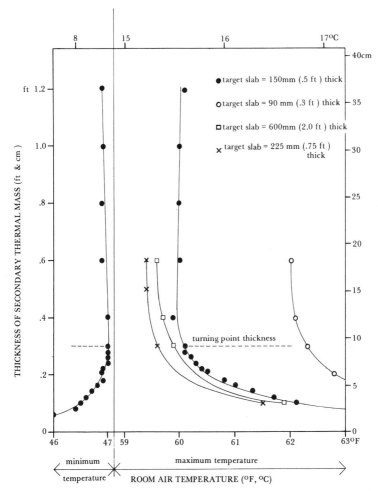

Fig. 3.16. The effect of the thickness of secondary thermal mass on the daily room air temperature fluctuation.

Among other things, the simulations revealed that the temperature of the surface of the secondary thermal mass lies somewhere between the temperature of the room air and the temperature of the surface of the target area, the target surface temperature being the hottest of the three during the charging mode.

The results drawn on the graph in Fig. 3.16 indicate several patterns of behaviour of passive solar buildings of this type. The area of the secondary

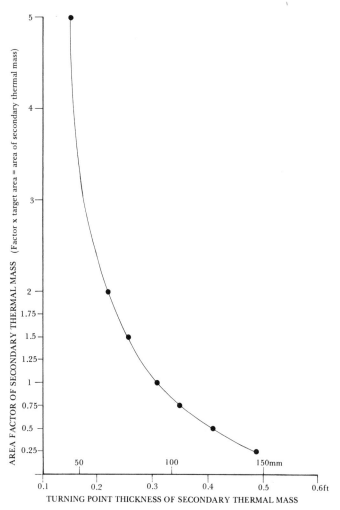

Fig. 3.17.   The relationship of the area of secondary thermal mass to its turning
point thickness.

thermal mass is maintained equal to the area of the target, and the areas of
these are not changed throughout the results shown on this graph. When
only the thickness of the secondary thermal mass layer is changed, it can be
seen that a thin layer will cause large daily air temperature fluctuations
within the room. As the thickness of the secondary thermal mass layer is
increased the air temperature fluctuation within the room reduces until a

point is reached where an increase in thickness will no longer reduce the air temperature fluctuation; this point is termed the turning point thickness.

Each of the subsidiary curves has been drawn using a different thickness of primary thermal mass (target) and again changing only the thickness of the secondary thermal mass layer. It can be seen from these curves that the turning point thickness of the secondary thermal mass is not greatly affected by a change in the thickness of the target slab. However, the turning point thickness was seen to change with a change in the insolation values from those listed in Table 3.3. A lower turning point thickness was obtained with a reduced insolation value. Furthermore, it is suspected that a change in values of specific heat, density and conductivity of both thermal masses will alter the value of the turning point thickness. Additional research is required in these areas.

A change in the proportional area of the secondary thermal mass, with all other parameters unaltered, will cause a change in the turning point thickness of the secondary thermal mass. The resulting changes in the turning point thickness are shown on the graph in Fig. 3.17.

Figure 3.18 indicates the changes which occur to the temperature fluctuations of the room air and target surface. These changes in temperature fluctuations result from changes in proportional areas of secondary thermal mass. The assumption in Fig. 3.18 is that when a proportional area of secondary thermal mass is modelled, the turning point thickness appropriate to that proportional area is also used for modelling. The two scales on the y–axis are intended to be read as one scale. Although possibly not feasible to construct, an area of secondary thermal mass five times the target area was modelled to reveal the trend of this graph. Such an area of secondary thermal mass has a turning point thickness of 4·5 cm (0·15 ft). The resulting air temperature fluctuation was 13·9–16·8 °C, approx. 3 °C (57·0–62·2 °F, approx. 5 °F). Thus it can be seen that although the amplitude of the daily room air temperature fluctuation is greatly reduced with an increased area of secondary thermal mass, the differences in maximum air temperatures produced are small. This indicates that the smaller secondary thermal mass area will provide the designer with smaller solar heating fractions (requiring more auxiliary heating).

It is interesting to see the degree of temperature control that the designer is capable of employing through the use of thermal mass. Such taming of the daily air temperature fluctuations would still be seen if less severe outdoor temperatures were used, and incidental gains included, resulting in room air temperatures within the comfort zone. This experiment was conducted and the results are shown in Fig. 3.19. In this set of

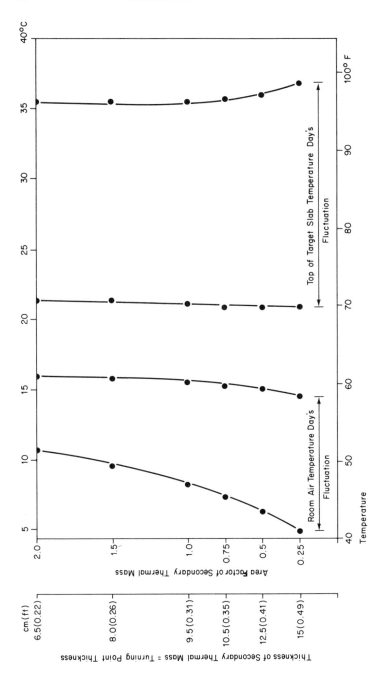

Fig. 3.18. Temperature fluctuations related to a change in linked area plus turning point thickness of secondary thermal mass. For each area of secondary thermal mass modelled, its turning point thickness is used. Noon target area × area factor of secondary thermal mass = area of secondary thermal mass.

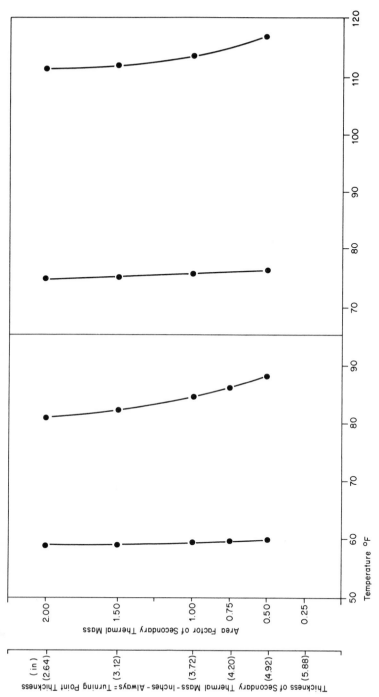

Fig. 3.19. Overheated example of data modelled in Fig. 3.18. Noon target area × area factor of secondary thermal mass = area of secondary thermal mass.

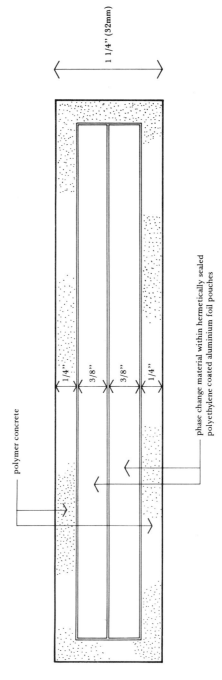

Fig. 3.20. A section through the phase change material tile.

polymer concrete

phase change material within hermetically sealed
polyethylene coated aluminium foil pouches

1/4"   3/8"   3/8"   1/4"

1 1/4" (32mm)

circumstances the minimum daily temperatures remained almost constant and the maximum daily temperatures were reduced as the area of secondary thermal mass was increased. The reduction of the maximum daily temperatures was found to be precisely the same as the increase in minimum daily temperatures shown on Fig. 3.18.

A third conclusion may be drawn from the information presented on the graph in Fig. 3.16: it was found that a thickness of approximately 225 mm (0·75 ft) was the optimum for a concrete target slab. This is in good agreement with the findings of Balcomb *et al.* [14].

Table 3.4.   *Heat capacities and thicknesses of various thermal mass alternatives*

Material	Concrete	Water	pcm Tile
Heat capacity	1 478 Wh/m^2 (476 Btu/ft^2)	1 478 Wh/m^2 (476 Btu/ft^2)	620 Wh/m^2 (220 Btu/ft^2)
Temperature swing	10 °C (20 °F)	10 °C (20 °F)	5·5 °C (10 °F)
Thickness	225 mm (0·75 ft)	115 mm (0·4 ft)	32 mm (0·1 ft (1$\frac{1}{4}$ in))

To reduce further the amplitude of the daily room air temperature fluctuations, it is necessary to turn to materials with higher heat capacities and conductivities than concrete. Table 3.4 indicates that water would be a good choice. This is especially so because of water's almost isothermal nature and the possibility of placing the absorber within this essentially transparent thermal mass. The disadvantage of water is that it has peculiar problems of containment and support. Another alternative, developed by Johnson, is the phase change material or pcm tile (Sol–Ar–Tile®) (Fig. 3.20).

With this tile, heat is stored latently at its freeze–thaw temperature. The phase change material employed is Glauber's Salts ($Na_2SO_4 \cdot 10H_2O$), which melts at 31·1 °C (88 °F), but is modified by additives to melt at 23·3 °C (74 °F). The thickness dimension of the tile is critical as the phase change material must be stored in layers no thicker than 9 mm ($\frac{3}{8}$ in) to prevent gravitational separation of the constituent materials. Thus the tile must be mounted horizontally. Polymer concrete is strong, light and waterproof and the surface finish can be varied. The tile weighs 5 kg/m^2 (11 lb/ft^2), only

® 'Sol–Ar–Tiles' are produced by the Cabot Corporation and the Architectural Research Corporation (ARC) in the USA.

slightly less than the weight of the same thermal capacity of water and the equivalent of one quarter of the weight of the same thermal capacity of concrete.

As described above, the comparative benefits of the pcm tile are significant, although not startling. The real value of the pcm tile comes to light when the control of room air temperature fluctuations is considered. The phase change of the material enables storage and release of energy with little change in tile core temperature (Fig. 3.21), which in effect provides a

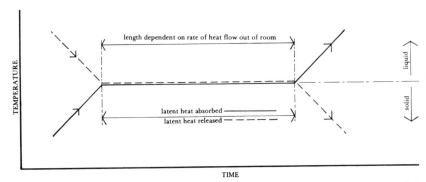

Fig. 3.21. The freeze–thaw temperature pattern of the phase change material (diagrammatic) [15].

built-in thermostat. Under steady state conditions, where the pcm tiles are the only source of heat, the energy flux from the core and the resulting time it takes for the core to freeze completely are variable, depending only on the total heat loss from the room to the outside. This is because the heat flow from the tile core into the room is, in such circumstances, equal to the heat flow from the room to the outside (Fig. 3.22).

Thus under these steady state conditions the air temperature of the room can be found from the relationship:

$$A_{store} \times U_{store} \times (T_{cor} - T_{rm}) = A_{ws} \times U_{ws} \times (T_{rm} - T_{amb}) \qquad (11)$$

where A = area, U = U–value, T = temperature.

Estimation of the air temperatures when the sun is shining into the tested room is more complex [16]. Computer simulations using the Sol–Ar–Tile® routines on Johnson's 'Thermal' program indicate that there is a clamping effect on the room air temperature fluctuation which approximately halves the fluctuation encountered with the use of a concrete slab target area, but the same daily minimum air temperature is maintained.

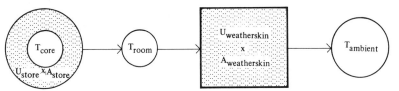

Fig. 3.22. A diagrammatic representation of the steady state energy flow from the Sol–Ar–Tile® [15].

Colloidal Materials Inc. (CMI)* have developed a heavy duty bag which will contain the phase change materials without the need for the polymer concrete casing. The cost of a 2 × 1 ft bag will be approximately $3·00 (£6·80/m²). This is then used with dense plaster board lining in the ceiling and is marketed under the tradename 'Heatpac'.

It is possible to construct a rule-of-thumb resistance split method for estimating the proportion of energy stored by a layer of thermal mass. This method is used in the passive design program (PDP) introduced in section 3.14 ('Overheating') and will be described here.

The rule-of-thumb method will simulate, with reasonable accuracy, the performance of a concrete target area thermal mass of thickness between 100 mm (4 in) and 225 mm (9 in), and a wood target area less than 50 mm (2 in) thick.

The energy absorbed by the target has two directions in which it is able to travel: into the air and surrounding objects by means of convection and radiation, and into the target slab by means of conduction.

The resistance of the air to energy flow in this direction is found as follows. Because the two paths into the air are in parallel, the U–values of each are added and the reciprocal of the result is found; this will be the resistance of the air to energy flow in this direction. This gives the same result as $R = (R_1 \times R_2)/(R_1 + R_2)$.

The resistance of the slab is found. When the concrete slab exceeds 175 mm (7 in) in thickness, the resistance is taken to be that of a 175 mm thick slab.

Then

$$Q_{st} = R_{air}/(R_{slab} + R_{air}) \qquad (12)$$

where Q_{st} = the fraction of the absorbed energy stored in the thermal mass,
R_{air} = the resistance to energy flow into the air (after absorption),
R_{slab} = the resistance to energy flow into the slab (after absorption).

* CMI, PO Box 696, Andover Massachusetts, USA.

Example. The target area slab is 150 mm (6 in) thick concrete. Assume the thermal coefficient to radiation is $5\cdot1$ W/m^2 °C ($0\cdot9$ Btu/h ft^2 °F); and the coefficient to convection is $2\cdot8$ W/m^2 °C ($0\cdot5$ Btu/h ft^2 °F). Assume also that the thermal resistance of 25 mm of concrete is $0\cdot018$ °C m^2/W ($0\cdot1$ °F ft^2 h/Btu).

Therefore the resistance of 150 mm of concrete $= 0\cdot11$ °C m^2/W ($0\cdot6$ °F ft^2 h/Btu) and the combined resistance to energy flow into the air $= 1/(5\cdot1 + 2\cdot8) = 0\cdot13$ °C m^2/W ($0\cdot71$ °F ft^2 h/Btu).

Then from eqn (12):

$$Q_{st} = 0\cdot13/(0\cdot11 + 0\cdot13) = 0\cdot54$$

Thus 54 % of the energy absorbed by the target slab will be stored there for later use, and the remaining 46 % will be lost by convection and radiation.

3.8.   CALCULATING SOLAR GAINS

Rules-of-Thumb
The quantity of useful solar energy collected by a south-facing vertical double-glazed window attached to a well-insulated house is as follows.

London: 140 k Wh/m^2 of window per year (48 000 Btu/ft^2)
Boston: 290 k Wh/m^2 of window per year (100 000 Btu/ft^2)

If a house is poorly insulated these figures will increase. Such figures are a useful guide with which to check more detailed calculations.

(a) Walls

The external surface colour of the opaque elements of a building's weather skin will determine the percent of the incident solar energy being absorbed. The subsequent movement of this energy through the element may be determined rigorously by using a finite difference thermal analysis program. The time lag involved in the conduction of the energy from the external to the internal surface will increase with an increase in the thermal capacity of the element. An increase in the thermal resistance of the element will result in a smaller quantity of energy transmitted. Thus in the passively heated systems normally encountered, where levels of external insulation are high, the solar energy transmitted has been reduced to such an extent that, for the purposes of heat load calculations, it can be neglected.

Where an opaque element of the weather skin has not been well insulated, the resistance split method* may be used to determine the quantity of

* This method is presented at the end of Section 3.7 ('Thermal Mass').

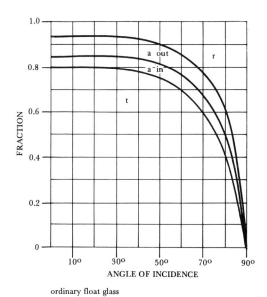

ordinary float glass

Fig. 3.23. Solar-optical properties of standard glass [17]. Above top line: reflectance (r), below lower line: transmission (t), between upper and lower lines: absorbance (a), up to middle line: solar gain factor (including transmission plus part of absorbed energy emitted inwards).

energy remaining within that element. All the energy remaining within that element can, by definition, be considered only as internal gains. If the element is shaded at any time during the day this should be taken into account.

(b) Windows

(i) *Information for heat load calculations.* As previously described, the proportion of solar energy passing through a transparent membrane is dependent on the transmission characteristics of the membrane to solar radiation and the hourly angle of incidence made by the sun on the membrane's surface. As the incident angle increases (Fig. 3.23) from $\theta = 0°$, the transmission to direct solar radiation diminishes, the reflectance increases and the absorptance initially increases because of the lengthened optical path and then decreases as the impact of the increased reflectance begins to take effect. The energy absorbed is divided equally between its two possible directions of escape: into the building and to the outside. Thus the heat gain factor is a percentage of the incident solar energy which includes both transmission and one-half of absorption.

The results of the solar interference boundary tests* are used to calculate the monthly total solar energy gains. Each window is taken separately if it is shaded at different times of the day, and it is analysed for each hour of the typical day of each month in the heating season. The typical day is usually taken to be the 21st of each month.

If an area of window is in shade for that hour, then the value for diffuse radiation and the heat gain factor for diffuse radiation are used. But if the area of window is potentially in sunshine, complications develop: rigorous analysis would involve separating the direct and diffuse portions of the energy received, for they have different heat gain factors; the mean percent diffuse radiation for each hour would be required, and the heat gain factor for the direct portion would require changing for each hour modelled. To avoid these complexities, an average heat gain factor can be developed for each month of the heating season and for each orientation being assessed. The average heat gain factor is applied only when the area of window is potentially seeing the sun.

Finding the average heat gain factor for each month is conceptually an easy task, but both this and the solar gain method presented below are difficult to perform without the use of a programmable calculator. The programs used will not be presented here because they are simply a translation of the equations provided.

The method used for computing the average heat gain factor is developed from eqn (8) (Section 3.5). Equation (8) is expanded to become eqn (13).

$$I_{tv} = \left(\frac{I_{dh}}{\cos \theta} \times \cos A \times \cos W \times T_{ad} \right) + (0{\cdot}5 \times I_{th} \times T_{af}) \qquad (13)$$

where T_{ad} = The heat gain factor for beam solar radiation for that hour, month and orientation; dependent on the properties of the transparent membrane and on the angle of incidence. Taken from a graph describing the membrane used (similar to that shown on Figs. 3.23 and 3.24).

T_{af} = The heat gain factor for diffuse solar radiation ($\simeq$ the transmittance of that membrane to beam radiation at an angle of incidence of $60°$).

Equation (13) is run for each hour of potential sunshine in the day being modelled. The sum of the results of this indicates how much energy is getting through the transparent membrane. The same runs are executed using eqn (8). The sum of the results of eqn (8) indicates the total quantity of

* See Section 3.4 ('Solar Interference Boundaries').

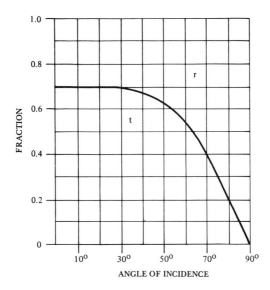

Fig. 3.24. Estimated transmission of a single layer of Suntek's heat mirror in relationship to angles of incidence. Above line: reflectance (r), below line: transmission (t).

energy incident on the surface of the transparent membrane for that day. The average heat gain factor is attained by dividing the total of the results from eqn (13) by the total of the results from eqn (8). Not only will the resulting heat gain factor differ for each month but the average heat gain factor for a south-facing window will differ from that for an east- or west-facing window.

The solar gains through each window can be computed by using eqn (14) for each hour of the typical day of each month in the heating season.

$$Q_{sg} = ((I_{tvdf} \times C_l \times A_d \times T_{adf}) + (I_{tvf} \times A_f \times T_{af})) \times M \times R_l \quad (14)$$

where Q_{sg} = the solar gain for that hour and window,

I_{tvdf} = the total (direct + diffuse) clear day insolation value for the vertical orientation concerned, for that hour; from tables [6, 7],

C_l = the climatic factor needed to convert I_{tvdf} into an average day insolation value for that hour—the method for finding this factor is outlined in Section 3.5 ('Solar Data'). Result of eqn (2), Section 3.5,

A_d = the area of the window in potential sunshine that hour,

T_{adf} = the average heat gain factor for the month and orientation considered; obtained from the method outlined above,

I_{tvf} = the quantity of diffuse solar energy falling on a vertical surface of the orientation, hour and month considered,

A_f = the area of the window in shade that hour,

T_{af} = the solar heat gain factor for diffuse solar energy,

M = a correction factor which includes an area correction factor to account for the area of window that is taken up by frames ($\simeq 0.85$) × a dirt factor,

R_l = a factor to account for the proportion of solar energy being back-reflected out through the windows.

In this equation R_l is found by estimating the number of reflections occurring, and the quantity of energy absorbed on each reflection, within the room before the light is back-reflected out through the window. It should be remembered that the solar energy will normally be reflected diffusely within a room. It is thought that this factor will vary from 0.6 in a room with large windows and light-coloured surfaces, resulting in 40% back losses, to 0.98 in a room with dark surfaces and small windows (or solar modulators), reducing back losses to only 2%.

The presence of dirt on the transparent membrane will reduce transmittance by 2–4% [70] depending on the frequency of window cleaning and on the presence of pollutants in the air. Thus the dirt factor will be between 0.96 and 0.98.

The results of eqn (14) are found for each hour, window and month of the heating season. They are summed to give the solar gain for the day and are multiplied by the number of days in the month being modelled. Finally they are divided by 1000 to give kWh gains per month (or by 1 000 000 to give MBtu per month).

An example of this data requirement is listed in Appendix E, Table E.19.

If one is considering a south-oriented collection surface, the method developed at Los Alamos Scientific Laboratory may be used [18]: The total monthly solar radiation on a horizontal surface is modified by the information shown on Fig. 3.25. 'Latitude minus mid-month solar declination' $(L - \delta)$ for each month is calculated, having first found the mid-month solar declination:

$$\delta = 23.3 \times \cos(30M_n - 187) \qquad (15)$$

where M_n = month number ($= 1$ for January, 12 for December) and where all values are expressed in degrees.

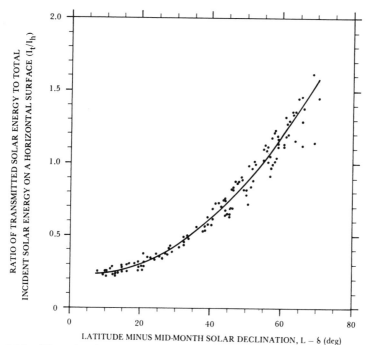

Fig. 3.25. The ratio of solar flux transmitted through a unit area of south-facing double-glazed window to solar flux incident on a horizontal surface versus latitude minus mid-month solar declination [18]. (The standard deviation is 0·060.)

The value of I_t/I_h for each of the monthly values of $(L - \delta)$ is determined, then I_t is calculated from the following formula:

$$I_t = (I_t/I_h) \times I_h \tag{16}$$

where I_t = the total monthly solar radiation transmitted through a unit area of a vertical south-facing double-glazing window.

(*ii*) *Information for the overheating program* (*PDP*).* The solar data required for this program are outlined in Section 3.5(b). The methods for deriving these data will be discussed in this section.

The insolation values found in the tables [6, 7] for vertical surfaces give the horizontal component of the incident solar energy (Fig. 3.26). PDP models the sunlight falling on a horizontal surface after passing through a vertical window; the program internally converts these horizontal components of the insolation values into their vertical components.

* Introduced in Section 3.14 ('Overheating').

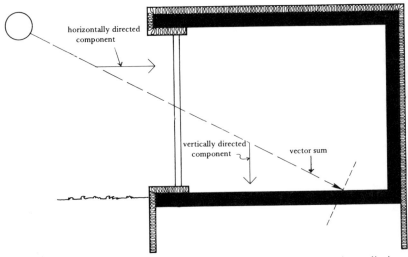

Fig. 3.26. Vertical and horizontal components of the incident solar radiation.

Corrections for transmission losses and for the losses due to the area of window frame and dirt are applied to the horizontal component value before entering the data into the calculator.

Equation (17) is used to correct the insolation values listed in the tables [6, 7].

$$Q_h = I_{hor} \times T_{ad} \times M \qquad (17)$$

where Q_h = hourly horizontal component of the solar energy gains on a clear day,

I_{hor} = hourly horizontal component of the clear day insolation value, taken from tables [6, 7],

T_{ad} = heat gain factor for that hour: transmission plus half the absorptance,

M = area factor × a dirt factor.

In eqn (17) T_{ad} is used only when tables show insolation values which are not corrected for the transmission through the membrane being used. T_{ad} is obtained from the graph of the solar-optical properties of the transparent membrane being used, allowance being made for the number of layers of transparent membrane. The angle of incidence, needed to find T_{ad}, is obtained from the Sun Angle Calculator®.

® Libby–Owens–Ford Co., Merchandising Dept., 811 Madison Ave., Toledo, Ohio 43695, USA.

In eqn (17) I_{hor} is the total (direct and diffuse) insolation value for that hour on a clear day. Because it is relatively small, the diffuse portion of the solar energy is treated as though it were direct radiation. Examples of these data are shown in Appendix E, Tables E.2 and E.3.

A second data requirement for PDP is the information which tells how much of the solar energy received by the collecting aperture is falling on the target.

For simplicity, the area of solar impact on the target is modelled by assuming that the building is facing due south and the month considered will be that of either equinox. These are large and constraining assumptions but they facilitate easy handling of this information. At the equinoxes (23 September and 21 March) for a building facing due south, the shadow or profile angle is constant throughout the day. Thus the depth of penetration of sunlight into the room for the whole day is determined by the profile angle at the equinox (Sun Angle Calculator®, or 90° − angle of latitude) and the height of the window (Fig. 3.27). The noon target area is then the product of the depth of penetration and the length of the window.

If at noon the area of sunshine extends up the rear wall, an effective area of target will have to be used for program data input (Fig. 3.28). Because the program uses the vertically directed component of solar gains (see calculations [19, 20]), it would be mathematically incorrect to use the combined horizontal and vertical area of target unless the profile angle is 45° (in which case the equivalent horizontal area of insolation is equal to the vertical area). Where the profile angle is not 45° the volume of the target remains the actual volume (i.e. using the vertical mass) and the area becomes the effective area (see Fig. 3.28).

If such a situation is very pronounced the results from PDP will not be reliable, because the effective area will be far greater than the actual area of the target and the program will allow for the thermal exchange over this larger area.

Knowing the solar azimuth for each hour of the day (Sun Angle Calculator®), it is possible to calculate the fraction of noon target area hit for each of these hours (Fig. 3.29). The fraction of the target area illuminated will be symmetrical about noon under these specified conditions.

It is assumed, within this program, that of the energy not falling on the target, approximately 50 % is going directly into increasing the temperature of the room air, the remaining 50 % being back-reflected out through the windows. The portion of this energy going directly into heating the room air is doing so because it is assumed that the surfaces other than the target area are reflective (off-white may be as low as 70 % reflective) and that, other than

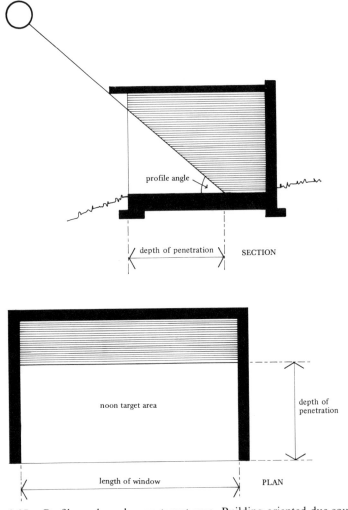

Fig. 3.27.   Profile angle and noon target area. Building oriented due south.

the secondary thermal mass area, all surfaces will have low thermal conductances and capacitances.

If the building being modelled is very long, the side wall effects are minimal and the fraction of noon target area hit will be 1·0 for all 8 hours. In any rectangular space, if the side walls are dark coloured and massive they act as primary thermal mass and again the fraction of target area hit will be

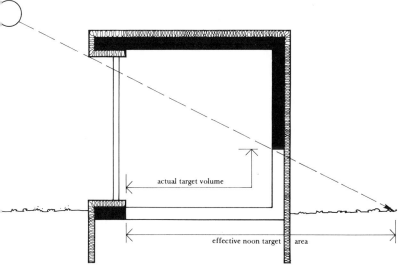

Fig. 3.28. An illustration of the effective target area.

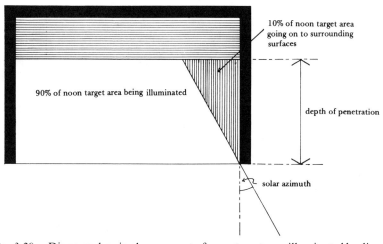

Fig. 3.29. Diagram showing how percent of noon target area illuminated by direct insolation varies throughout the day due to the changing solar azimuth angle. Building orientation due south, at time of equinox.

1·0 for all 8 hours. An example of this data is listed in Appendix E, Table E.1.

As mentioned above, the method adopted within PDP is to calculate the solar energy gains by finding the product of the area of target being illuminated and the vertical component of the hourly insolation gains per unit area. If the collection aperture does not face due south or if the month being modelled is not the equinox, using PDP becomes more complex and it is advised that one uses the thermal network analysis program (PSP).*

A point to consider carefully is that the solar radiation should not fall on to furnishings, or, if it does, that it must be accounted for in the percent of noon target area illuminated.

3.9. THE SOLAR MODULATOR†

A diagrammatic representation of the use of the solar modulator is shown in Fig. 2.18, ('winter day' of system schematics), Section 2.3.

In directly heated examples of passive solar buildings there is a natural tendency to use the floor as the primary thermal storage area. This causes conflicts between solar energy collection and space use. Furnishings within this area will be the cause of two unwanted effects: they will conceal thermal mass‡ and thus increase the quantity of energy going directly into the room air, and, being both flooded with light and visually detailed, they will be the cause of contrast glare. Glare within this type of passive solar building is also caused by the need for large areas of unobstructed sky view.‡‡

The solar modulator reduces these conflicts to a minimum by reflecting direct solar energy on to the ceiling. Thus the ceiling, which is usually free from space use requirements, becomes the primary thermal storage area. The development of such a system goes hand-in-hand with the development of the phase change material tile (Sol–Ar–Tile®). This tile, due to its economy in weight, may be applied to the ceiling in place of a normal ceiling finish. By reflecting solar radiation on to the ceiling, the solar modulator reduces glare to a minimum in three ways: by partially obscuring the glare source, by removing the possibility of contrast glare from the field of vision

* PSP (passive simulation program) is introduced in Section 3.14 ('Overheating')
† Conceived and developed at MIT by Timothy E. Johnson and Dennis A. Andrejko [21].
‡ See Section 3.7 ('Thermal Mass').
‡‡ See Section 3.19 ('Glare') for discussion on types of glare.

nd by drawing much of the light into the back of the room, resulting in igher levels of surround illumination there.

The solar modulator has four additional advantages: it does not obscure r detract from the view while reflecting solar energy on to the ceiling, it will erve the same purpose as drapes or curtains, providing privacy at night, it vill reduce fading of furnishings due to sunlight, and it will reflect unwanted olar energy away from the building in summer, again without obscuring he view.

The concept of the solar modulator is that the slats of a standard venetian lind are inverted and a reflective finish is applied to the top surface of each lat. Unfortunately the practical solution is not so simple.

The constraints imposed on the design of the optimum reflective louvre re numerous. These constraints are listed here in an attempt to formulate uidelines for an architecturally acceptable product.

(a) Sunlight should be reflected off the solar modulator in such a way that does not cause blinding glare to the occupants of the room. This requires oth architectural and louvre design constraints. A reasonable constraint n the louvre is to stipulate that the angle subtended by that beam, which fter reflection is closest to horizontal, and a line parallel to the floor, be no maller than $30°$.

(b) The louvre would be enclosed within the south-oriented glazing unit o prevent reduced reflectance due to accumulation of dust or atmospheric ollution. This introduces two further constraints. The materials from vhich the louvre is made must be able to withstand heat build-up in the nterpane space, otherwise sagging or distortion could result. The angle vhich in (a) is limited to a minimum of $30°$ should also be limited to a naximum of $72°$ to prevent excessive reflection losses at the outer surface of he inner layer of glass.

(c) The solar modulator should intercept 100% of the incident solar nergy during the central hours of the day. Ideally all solar energy hroughout the day should be intercepted, but this proved to be impossible o achieve without adjustment and without infringing on some of the other ore critical constraints. Thus it was decided that between the hours of .30 a.m. and 2.30 p.m. the system should work within this constraint.

(d) None of the reflected sunlight should be intercepted by the above slat.

(e) The slats should never have to be tilted to such an extent that they etract from the view. This subjective constraint should be experimentally efined.

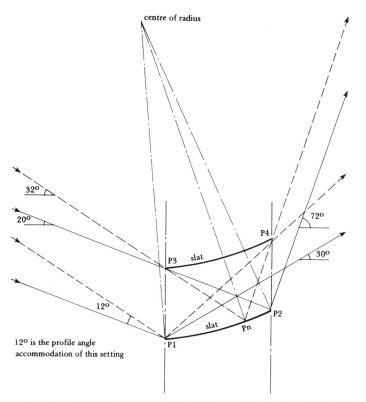

Fig. 3.30. Ray diagram analysis of the solar modulator for the first setting of the season (→ 21 December →) at 40°N latitude.

(f) The horizontal alignment of the slats is critical and cannot be allowed to change with the age of the louvre. Stretching of the supports for the louvre could result in constraint (a) being met at the top of the louvre but slats towards the bottom of the louvre reflecting at angles shallower than 30°. This is the case with the initial louvres of the MIT Solar 5 building. Research into the use of a non-stretch material, such as Kevlar®, is being done as a result of such findings.

(g) To avoid excessive 'babysitting' of the system, a minimum number of seasonal adjustments is aimed for. The radius of curvature of the slats is a critical dimension which, together with the spacing between the slats, determines the number of times the angle of the slats requires adjustment each season.

Translating some of these constraints is made simple by analysing Fig. 3.30. P_1 and P_2 are the outer and inner points of a slat within a reflective louvre placed at 40 °N latitude. Rays reflected off P_1, P_n or P_2 must not be greater than 72 °, nor less than 30 °, nor hit the slat above. Such constraints determine the maximum profile angle range accommodated by the louvre without a change in the tilt of the slats. The profile angle is discussed in detail below.

Modelling

The (apparent) movement of the sun across the sky is complex to model because the path the sun traces changes every day.

In the northern hemisphere, if one faces due south the sun will be seen to trace a path closest to the horizon on 21 December and furthest from the horizon on 21 June (the two solstices). The only way to make any sense of all this movement is to use what is called the profile angle (see Fig. 3.31) and to model the louvres as facing south. The profile angle is a measure of the altitude of the sun, but it is corrected so that the angle being described is perpendicular to the window or edge of the louvre, whereas the true altitude of the sun is taken along the solar azimuth (angle in plan between due south and the sun). They coincide only at solar noon of every day for a window facing due south.

Requiring the orientation to be due south is necessary for proper functioning of the louvre. It also means that the solar profile angles are symmetrical about solar noon each day, and about 21 December each year. Thus only half a day and half the heating season need be modelled. For this reason all results concerning the number of slat adjustments during the heating season are for only half the heating season ($\rightarrow$ 21 December, or 21 December $\rightarrow$). To obtain the number of adjustments required for the whole heating season, the figures presented in this section should be doubled first and then one adjustment subtracted. One adjustment is subtracted because the first setting, for 21 December, is symmetrical about this date and thus requires only one setting from before and to after 21 December.

The first look at the information on Figs. 3.30 and 3.31 would lead one to suspect that there would be no difficulty getting through the heating season if the first setting accommodates a 12 ° profile angle range and the profile angle range of the total heating season is 30 ° (20 ° start at 9.30 a.m. on 21 December and 50 ° + at the end of the heating season). The difficulty arises because the end of the usefulness of one setting is determined by the noon profile angle but the next setting must reflect the next morning's 9.30 a.m. sunlight perfectly. Except at the equinoxes and between (in summer), this

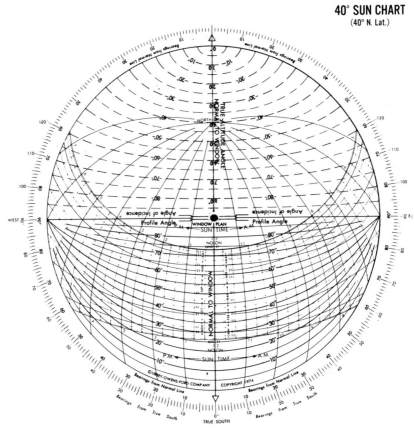

Fig. 3.31. Sun Angle Calculator® (by courtesy of Libbey–Owens–Ford Co.)[22].

morning profile angle is noticeably smaller than the noon profile angle of the preceding day. Thus the number of settings for the heating season will be considerably more than one might at first anticipate.

Testing

The slat radii of most standard venetian blinds are too small to accommodate a large range of solar profile angles without adjustment. Thus an optimum radius must be sought. There are three methods for developing and testing the optimum slat radius and spacing between slats: graphic testing using ray diagrams, a model testing rig, and computer simulation.

Ray diagrams such as Fig. 3.30 are drawn with different slat settings. The

Fig. 3.32. The solar modulator test rig.

lowest profile angle of the heating season occurs on the morning of 21 December. This, together with maintaining a minimum reflected angle of 30° within the room, determines the maximum spacing between the slats.

The testing rig used is shown in Figs 3.32 and 3.33. The rig shown here does not give as accurate results as the ray diagram method, but it is certainly possible to construct a rig with such precision.

Fig. 3.33. The solar modulator test rig (shows part of the profile angle indicator).

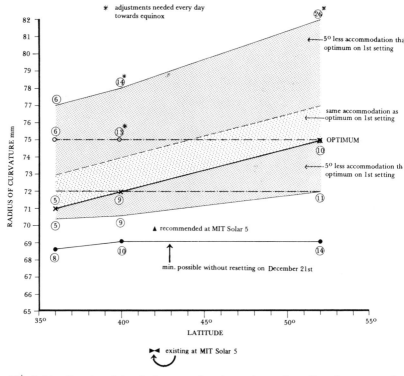

Fig. 3.34. Results of simulation tests showing a change in radius of curvature for each of three modelled latitudes. A newly determined slat separation is used for each point. Encircled numbers represent the number of adjustments needed for the heating season (from 21 December on).

Both methods have their advantages. The testing rig gives very quick results but requires an actual slat to model. Thus only existing slat radii may be tested. In contrast, the ray diagram method is laborious but versatile.

Computer simulation was carried out on the TI–59 (Texas Instruments) programmable calculator. This was more accurate and less cumbersome than the ray diagrams. Having achieved initial agreement with the results from ray diagrams for 40 °N latitude, an attempt was made to cover a range of latitudes which would encompass the whole of the USA and most of Europe. Two objectives were in mind: to determine if the optimum radius of curvature was different for different latitudes, and to investigate the extent of slat separation change due to the change in latitude.

The results of this testing are shown in Figs 3.34 to 3.36. In Fig. 3.34 a

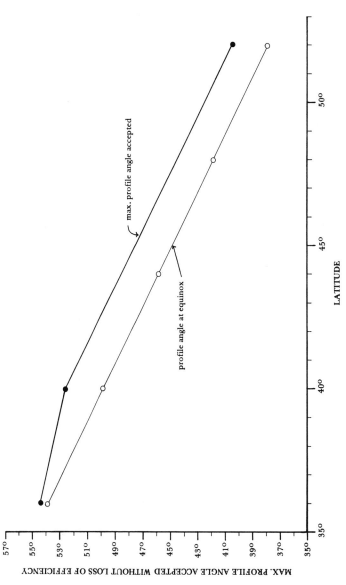

Fig. 3.35. Maximum profile angle perfectly reflected at each of three latitudes compared with the equinox profile angles. Minimum possible radius of curvature (Fig. 3.34) is used.

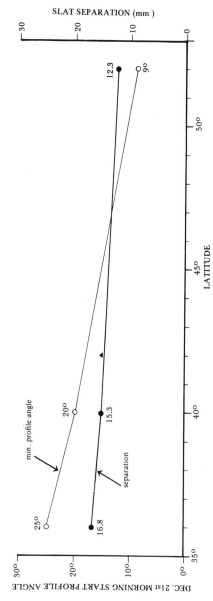

Fig. 3.36. Results of the change in slat separation with change in latitude. Minimum possible radius of curvature (Fig. 3.34) is used.

newly determined slat separation is used for each simulation of either a change in radius of curvature or a change in latitude. The optimum radius of curvature is seen to change from 71 to 72 to 75 mm for a 25 mm wide slat at latitudes of 36°, 40° and 52° respectively. The number of slat-pivoting adjustments required through the heating season (21 December → just past equinox) is encircled for each point modelled. Inverted Rolscreen® venetian blinds* were the only commercially available louvres which came anywhere near the radius of curvature requirements of the MIT Solar 5 building. It does fall below the minimum possible radius of curvature to make it through 21 December without midday adjustments (Fig. 3.34). Some complaints of glare are suspected to be caused by this. In terms of the number of adjustments, the reflective louvre is seen to work best at lower latitudes. Manual adjustment of the louvre at these low latitudes is acceptable (nine adjustments in six months), but at a latitude of 52° such manual adjustment may not be acceptable (19 adjustments in six months or one every ten days). As a rule the frequency of adjustments necessary is greater towards the equinox and fewer towards the winter solstice (21 December).

The minimum possible radius of curvature for each latitude (from Fig. 3.34) was used to draw the information on Figs 3.35 and 3.36. Figure 3.35 indicates the maximum profile angle being perfectly reflected under such circumstances. 'Perfectly' here means not infringing on any of the design constraints outlined above. It was found that this maximum profile angle (Fig. 3.35) was reduced by only a few degrees when the optimum radius (with its associated slat separation) is modelled, but that it does not fall below the equinox profile angle (Fig. 3.35). In some locations the heating season can extend well past the equinox into April and May. This present lack of extendability of the reflective louvre, making it partially inoperable on underheated days well beyond the equinox, may prove to be an important limitation. For at this time of year (March, April) the tendency of the living space to overheat will be increased if improper functioning of the louvre occurs.

Figure 3.36 shows the maximum slat separation possible for a 25 mm wide slat at each latitude. The slat separation changes with both the radius of curvature and the latitude.

It is clearly not commercially feasible to produce a ladder (this controls slat separation) nor a radius of curvature for each change in latitude. Therefore the next task, which remains somewhat incomplete, was to

* By the Rollscreen Company, Pella, Iowa 50219, USA.

determine the sensitivity of the design constraints to these two parameters (radius and separation).

Figure 3.34 indicates that the system is not extremely sensitive to radius of curvature change (± 1 mm). This is fortunate because one of the problems of manufacturing aluminium venetian blind slats is that it is almost impossible to predict the final radius of curvature produced by a rolling die. This is because the slat radius will change after the slat is rolled. Thus only by trial and error is it possible to produce the correct radius. It is an expensive process because of the high costs of the die. On Fig. 3.34 the horizontal band enclosed by the broken lines is a range of radii which would be acceptably well suited to any of the range of latitudes in the USA and most of Europe. It is thought that a 72 mm radius of curvature for a 25 mm wide slat will perform best over all these latitudes.

Using such an optimum slat radius, it was necessary to test if a fixed ladder dimension could be used for lattitudes above and below the one for which the ladder was perfected. These tests revealed a new point of departure: it was not possible to move down latitude but extremely advantageous to move up latitude. For example, the ladder for 36° latitude could be used at 40° latitude, with the identical radius of curvature, and needed only five adjustments for the season (compared with the previously found optimum of nine—Fig. 3.34). It was here realized that the slat separation set by the morning profile angle on 21 December is only the *maximum* slat separation and perhaps one could attain better results by reducing this slat separation. There are two possible pitfalls to this. The angles reflected off the louvre in such conditions tend to be high and thus wasteful of energy, and the slats will have to be tilted more under these conditions, with the possibility that the view may be unacceptably obscured. It is suspected from observing one test run that changing the radius of curvature may counteract both pitfalls to some extent.

The problem of not being able to reflect sunlight perfectly beyond the equinox may be overcome by doubling the separation between the slats when approaching 21 March. This was realized when modelling showed P_4 (Fig. 3.30) to be intercepting reflected sunlight (from the slat below) towards the equinox and thus preventing large accommodation angles. Doubling the slat spacing could be achieved by only two additional strings in the louvre system (one at either end) which would be pulled to increase spacing (Fig. 3.37). Each string is attached to every second slat but is free to move within the above slat. The result of this doubling at 40 °N latitude gave a maximum possible perfectly reflected profile angle of 67·5° (towards the end of April) with five additional adjustments from the equinox.

The next step in this research is to test these possibilities fully. To do so will require the use of a larger capacity computer with the capabilities of higher speeds and graphic display.

It is conceivable that, with a combination of these untested alternatives, a louvre system could be developed which will allow perfect reflection of solar energy throughout the heating season, with an acceptable number of adjustments, for all latitudes in the USA and for most latitudes in Europe.

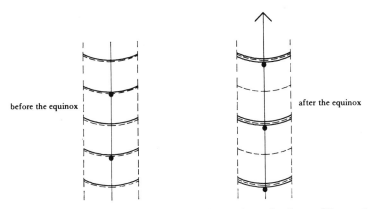

before the equinox   after the equinox

Fig. 3.37.   Doubled louvre spacing to allow for perfect reflection well beyond the equinox. Patent pending.

It is possible to convert the values in this section so that they are applicable to any desired slat width, given that:

$$\frac{S_a}{S_b} = \frac{s_a}{s_b} = \frac{r_a}{r_b} \tag{18}$$

where S = slat width, s = separation of slats, and r = radius of curvature of slat.

3.10.   INCIDENTAL HEAT GAINS

To shorten this section, solar gains have been dealt with separately in Section 3.8. The incidental gains referred to in this section do not therefore include solar gains.

In a passively solar-heated row house the incidental heat gains will commonly reduce the resulting heating load by 60%, the remaining

Table 3.5. *Average daily incidental heat gains from inhabitants, appliances and hot water for a three-bedroom town or row house in winter (kWh/day, Btu/day)*

Source Units	EUROPE										USA	
	Siviour [24]		Billington [25]		Wolf [26]		Brundrett [27]		Heap [28]		Hittman [23]	
	kWh	Btu	kWh	Btu	kWh	Btu	kWh	Btu	kWh	Btu	kWh	Btu
Occupants	5	17100	4·8	16400	5·4	18400	6	20500	4	13650	—	—
Lighting	4	13650	1·8	6100	12	41000	12	41000	1·4	4800	5	17100
Cooking	4	13650	8·0	27300					3·4	11600	6	20500
Appliances	4	13650	7·8	26600					3·4	11600	14	47800
TOTAL	17	58050	22	76400	17·4	59400	18	61500	12·2	41650	—	—
From hot water	5	17100	15	51200	—	—	11	37500	3·4	11600	—	—

30–40 % being carried by solar gains. In view of this it is surprising that so little research has been done on incidental heat gains in the United States.

Apart from the Hittman Report [23], which provides little of the required information (Table 3.5), the only documented research available is analyses of utility company statistics. These analyses do not provide sufficiently reliable data bases to work from, for they will not reveal to what extent the hot water heating is contributing to space heating, nor whether the clothes dryer or range (cooker) is vented to the outdoor air.

In Europe there has been much recent work investigating the magnitude of incidental heat gains in dwellings (Table 3.5). The results are much the same as those reported for American housing, with the exception that there is heavier use of electric appliances in the American home.

Unless appliances are vented to the outdoor air, their total electricity consumption is converted to incidental heat gains.

The energy supplied from the body heat of the occupants is broken down by Siviour [24] thus:

Two adults:	100 W each (day-time)	80 W each (night-time)
Two children:	60 W each (day-time)	50 W each (night-time)

8 h overnight:	$(2 \times 80) + (2 \times 50)$ W/h	2·08 kWh
1 h morning:	$(2 \times 100) + (2 \times 60)$ W/h	0·32 kWh
6 h daytime:	$(1 \times 100) + (2 \times 60)$ W/h	1·32 kWh
2 h afternoon:	vacant	
5 h evening:	$(2 \times 100) + (2 \times 60)$ W/h	1·60 kWh
2 h evening:	vacant	

5·32 kWh/day
= 18 150 Btu/day

One significant point about the incidental heat gains is that they vary throughout the year (Table 3.6). This variation will have repercussions on the balance point temperature* of the building and thus its occurrence cannot be overlooked.

The variation in incidental heat gains throughout the day for a typical English home is shown in Fig. 3.38. This type of information, although it does not account for all heat gains within the house, can be used as a basis for the data needed in the overheating design method program (PDP) in Section 3.14 ('Overheating'). Examples of the data requirement for the worked example are shown in Appendix E, Tables E.4 and E.22.

* See Section 3.11 ('Heating Load Calculations').

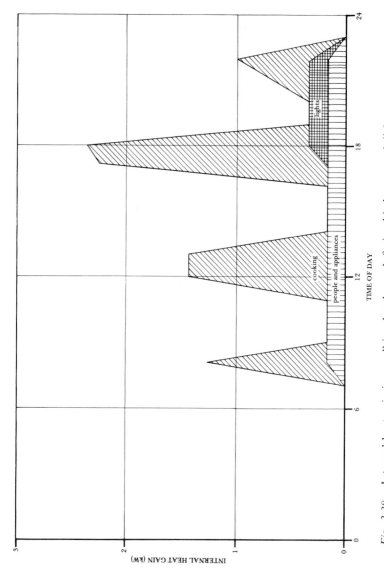

Fig. 3.38. Internal heat gain in a well-insulated, north-facing kitchen on a dull January day [29].

Table 3.6.   Incidental heat gains (kWh/day) for a typical three-bedroomed house in England [24]

Month	Appliances, lights and occupants	Hot water
January	19	5·0
February	19	5·0
March	18	4·5
April	16	4·0
May	14	3·5
June	12	3·0
July	12	3·0
August	14	3·5
September	16	4·0
October	18	4·5
November	19	5·0
December	19	5·0

3.11.   HEATING LOAD CALCULATIONS

Rules-of-Thumb

$$England: 1 \ W/°C \simeq 42 \ kWh/y$$
$$Useful \ incidental \ gains \simeq 5000 \ kWh/y$$

It is suggested that a figure of 5000 kWh/y for incidental heat gains should be used (people and appliance gains) for an average five-person household. The heat losses per season will vary considerably throughout the world and it is suggested that the reader constructs his or her own rules-of-thumb from empirical data. In England a reduction of total heat loss of the dwelling of 1 W/°C will give a variously reported reduction in annual heating load of between 40 and 50 kWh/y [30, 31]. An average figure is about 42 kWh/y.

From a logical point of view, reducing the overall heat loss (W/°C) of a dwelling should reduce the annual heating bill of that dwelling. This is not necessarily so; usually in England it will lead to increased thermal comfort up to a certain point and only then to fuel savings. These are inhabitant behaviour issues which, although important, are not quantifiable for design purposes. Nor is it advisable to suggest the relative weightings we should give to either economy of fuel use or thermal comfort.

It is interesting to note that there are no accurate methods of annual heating load calculations available. Some methods are mathematically more accurate than others, but with random sample field tests of identical

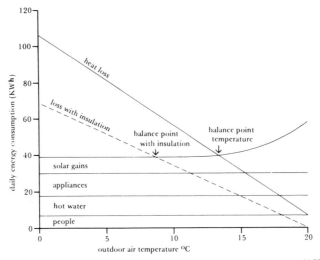

Fig. 3.39.   Illustration of the balance point temperature [27].

houses there are ratios of 4:1 in annual energy consumption [32]. This is due to inhabitant behaviour patterns and they cannot be predetermined when designing a dwelling. There are also other inaccuracies involved: in England the average monthly indoor temperatures are seen to change throughout the heating season; standard and passive solar buildings will probably have different average monthly indoor temperatures [33]. Thus it should be understood that both the rules-of-thumb above and the heating load calculations discussed are for comparing a standard house with a passive solar house, both at approximately the same average indoor temperatures during the heating season. More precise hourly heat load calculations may solve some of these problems but not all; the behavioural use problem will still exist.

The conventional methods of conducting heating load calculations require some consideration.

A diagrammatic representation of the effects of internal gains and insulation on the balance point temperature of a normal modern duplex (semi-detached house) in England is shown in Fig. 3.39.

The balance point temperature is the outdoor temperature above which no heating is required in the house. It is the point where the internal gains meet the thermal losses. The heating is not used until the outdoor temperature falls below this balance point. Thus it is the balance point temperature which determines the length of the heating season. Increased insulation, larger internal gains or lower infiltration rates will lower this balance point temperature.

Conventional methods of calculating heating loads mostly compute the fabric and infiltration heat losses due to the mean temperature differential and do not as accurately account for the length of the heating season. Another conventional method employs the use of published degree day figures for each month. In America this makes the assumptions that the balance point temperature (degree day base) is 18·3 °C (65 °F) and the thermostat setting is 21·1 °C (70 °F). Furthermore, the assumption is made on past performance that the 2·8 °C (5 °F) difference is contributed by the incidental gains. This past performance is that of single family residences, and because the balance point temperature is so sensitive to the thermal properties of the building (see Fig. 3.39) the use of this method for energy-conserving buildings will lead to erroneous results (see Appendix A).

In England the commonly accepted degree day base temperature is 15·5 °C (60 °F). This does not reflect the high insulation levels used in British housing but rather the low levels of comfort accepted there. It may also be partly due to the high level of water vapour usually present in the air during winter in Britain.

The Royal Institute of British Architects (RIBA) have designed a heating load calculation package for the TI–59 (Texas Instruments) programmable calculator which is based on an excellent method developed by the Joint Fuel Industries. This method is outstanding because it was developed from empirical data. The RIBA program has reduced the usefulness of this method by assuming that the contribution of incidental gains to offset the heating load extends for a full heating season of 243 days (8 months). The result of this assumption is that an energy conscious design can have a substantial negative heating load by simply reducing the length of the heating season. As seen from the discussion on balance point temperatures above, this is, in the Author's opinion, a misconception.

It is to the degree day method that we turn in an attempt to find a more realistic method for determining the heating load of a building. The equation which governs the balance point temperature is as follows:

$$T_{DD} = T_{th} - (Q_i/Q_1) \qquad (19)$$

where T_{DD} = the balance point or degree day base temperature,
   T_{th} = the average thermostat set point temperature,
   Q_i = the incidental energy gains per day due to people, appliances and solar energy,
   Q_1 = the heat loss per day °C (°F).

The heat loss per day °C

$$Q_1 = ((A_{ch} \times V \times P) + (U_{ws} \times A_{ws})) \times H \times dT \times 0.84 \qquad (20)$$

where A_{ch} = the number of air changes per hour,

V = the volume of air in the building,

P = the heat capacity (specific heat × density) of air
   = $0.34 \, \text{Wh/m}^3 \, °C \, (0.018 \, \text{Btu/h ft}^3 \, °F)$,

A_{ws} = the area of the weather skin,

H = the number of hours in the day = 24,

dT = internal → external temperature difference = $1 \, °C \, (1 \, °F)$,

0.84 = the correction factor according to Hittman Associates [34],

and where

$$U_{ws} = ((U_1 \times A_1) + (U_2 \times A_2) + \cdots + (U_n \times A_n))/A_{ws} \qquad (21)$$

where U = U–value (W/m^2 °C, Btu/h ft^2 °F),

A = area.

Because the internal gain per day varies each month (because of the sun's altitude, percent sunshine, and inhabitant behaviour patterns), the internal gain is dependent on the degree day base temperature. But the degree day base temperature is in turn dependent on the internal gains per day figure. Thus iteration of eqn (19) is necessary to narrow down on the correct result.

So by knowing the values for the variables on the right hand side of eqn (19) it is possible to calculate the balance point temperature for any building. And by knowing this temperature, approximate heating degree day figures for each month, for any building, can be determined thus:

$$DD = (T_{DD} - T_o) \times N_d \qquad (22)$$

where DD = degree days per month

T_o = mean monthly outdoor temperature

N_d = number of days per month.

If the result of $(T_{DD} - T_o)$ is less than zero, then the resulting negative heating load is discounted. Separate calculations are required to determine cooling load if any.

Then

$$Q_m = DD \times Q_1 \qquad (23)$$

where Q_m = monthly heating load.

To explain briefly the meaning of a degree day, if the balance point temperature, or degree day base temperature, of a building (eqn (19)) is 13 °C (55·4 °F) and the mean outdoor temperature for one day is 12 °C (53·6 °F), the number of degree days during this day is 1 (or 1·8).

The TI–59 program for this method of heating load calculation is described in Section 3.18 ('Cost Analysis'). In this program the balance point or degree day base temperature is determined by the methods set out above. However, the method of computing the number of degree days in eqn (22) does not allow for the few warm days in, say, November or the few cold days in, say, October. The costing program incorporates the more accurate 'bin' method of heating load calculation. This does account for those warm and cold days in the data used for determining the number of degree days. This gives the program results added accuracy.

The number of days per heating season of each 1 °C (or 2 °F) difference in mean daily temperature is found and the number of degree days produced by these is tabulated (as in Table 3.10 in Section 3.18).

For example, if the data obtained are as follows (brief extracts only):

Mean daily temperature (°C)	Number of days/ heating season
0	9
−1	5
−2	3
−3	3
−4	1
−5	1
−6	0

Then, if − 5 °C is the balance point temperature, the number of degree days is 0. At −4 °C balance point the number of degree days is 1. Similarly:

Balance point temperature	Number of degree days
−3 °C	3
−2 °C	8
−1 °C	16
0 °C	29
1 °C	51

The algorithm for doing this is as follows: Keep a running total of the number of degree days. Assume we are attempting to find the number of degree days at a balance point temperature $x°$. Increase the number of degree days at temperature $(x - 1)°$ by the number of days at the temperature $(x - 1)°$, plus the summed total of the number of days at all temperatures at and below $(x - 2)°$. The result is the number of degree days at a balance point temperature of $x°$. The full table of such figures for London is shown in Table 3.10 in Section 3.18.

One of the results of the Cost Benefit program is the balance point temperature for the building modelled. From knowing the balance point temperature, the corresponding number of degree days may be found from the table and the resulting heating load is computed using eqn (23).

Appendix A contains a sample degree day table for Boston, Mass., USA. This was computed by the author using weather tapes from the National Climatic Center, North Carolina.*

An alternative method of displaying the degree day data shown in Table 3.10 is to consider the varying balance point temperatures on a monthly basis. This would have two advantages: the heating load over the year could be graphically displayed; and if known one could vary the balance point temperature for each month to bring added accuracy, in line with observations by Chapman [33].

Research into ways of producing normal degree day data below any base temperature has been conducted by Thom, who has developed a method of arriving at such data from knowledge of monthly mean temperatures and the standard deviations of monthly mean temperatures [35, 36].

The reader is recommended to keep in touch with the now constant flow of thorough research from the Los Alamos Scientific Laboratory.† Passive solar system performance methods have been developed there but are not included with this handbook.‡ The following are some observations concerning the LASL Solar Load Ratio [18, 37] passive solar performance method.

(a) The way of handling system performance in this method is in an 'after-the-event' manner without providing the designer with an insight into comfort levels achieved by the passive system. The prime concern in passive

---

* A complete set of heating degree day tables, one for each of 379 weather stations in the USA, is available from the author.
† LASL, University of California, Los Alamos, New Mexico 87545, USA.
‡ The LASL method for Trombe–Michel and water walls has been transformed into a program for the TI–59 (Texas Instruments) programmable calculator plus printer by Fuller Moore. Cards and manual are available from Fuller Moore at a cost of $10 plus 6 replacement cards. Write to: Fuller Moore, R. A., Associate Professor of Architecture, Miami University, Oxford, Ohio 45056, USA.

olar design is that of comfort coupled with avoidance of waste of any potentially collectable energy. To these ends the Solar Load Ratio method provides little assistance.

(b) The thermal network analysis model used at LASL to produce this design method does not separate convective and radiative heat transfer from the surface of the thermal mass. The record of good agreement with empirical data from the test cells there is largely a result of there being no secondary thermal mass in the test cells—a situation untypical and undesirable in a passive solar system. Thus changes to the daily room air temperature fluctuation (seen in Fig. 3.16 in Section 3.7) and the resulting changes in seasonal solar heating fraction are not accounted for in the estimating method developed by LASL. In essence this, together with the observation made in (a), means that the hands of the designer are tied so that he cannot determine the effect of increasing or decreasing the area or thermal capacities of primary or secondary thermal mass, nor the effect of the weather (in different locations) on the comfort conditions within the building.

(c) The extent of daily temperature fluctuations allowed in this method is dictated. It is assumed that the air temperature be allowed to fluctuate $\pm 3\,°C$ (5 °F) around 20 °C (68 °F). If the air temperature exceeds 23 °C (73 °F), outdoor air ventilation will start and if the temperature drops below 17 °C (63 °F) the heating will be turned on. These limits cannot be applied to every design situation. Some clients will accept a larger temperature fluctuation around a lower mid-point which would increase the solar heating fraction performance.

(d) The Solar Load Ratio method was developed for the Trombe–Michel wall and was borrowed for direct gain systems, for which it is less appropriate. The concept behind the graphs in this method is the assumption that when the SLR fractions are between 1 and 2·5 there is much overheating which causes waste of potentially collectable solar energy through venting. This overheating effect can be designed out of direct gain systems by knowing the appropriate design methods. Thus the SLR method is not useful for direct gain systems.

(e) These methods require further validation for locations within the United States, and even more so for countries outside the USA, for it is well known that the comfort demands of Americans differ considerably from those of Europeans. Validation attempted by researchers at the Centre National de la Recherche Scientifique (CNRS), Odeillo, France, did not obtain good agreement between the results of the LASL method and the prototype houses there.

3.12.   FABRIC LOSSES

The average thermal resistance of the building's weather skin must be made as high as is economically feasible. Testing such economic feasibility is possible by means of the design method or program listed in Section 3.18 ('Cost Analysis'), and not by any intuitive decision.

The weather skin components of the building will be discussed in turn.

(a) *Windows*
These were discussed in Section 3.6 ('Windows: Improvement for Solar Collection').

(b) *Walls*
Insulation should be placed on the external surface of masonry walls. Doing so will bring them within the insulating envelope and allow them to perform as either primary or secondary thermal mass.* Calculations must be conducted to determine if and where condensation occurs and, if necessary, preventive measures taken. The method for determining the likelihood of condensation occurring within the building fabric is outlined elsewhere [38]. The presence of water within the fabric of the weather skin will both gradually destroy most building materials and considerably reduce their thermal resistance. Table 3.7 shows how the thermal conductivity of masonry walls will vary with exposure to rain. In conditions of driving rain [39] and where condensation occurs, even higher moisture contents can be expected [40]. If a house is seldom occupied it may be more cost effective to insulate internally, giving quick response time to occasional heating.

(c) *Roofs*
In most houses on the market, at least 25% of the total seasonal heating load is being lost through the roof. Remedial measures (insulation) are easy and cost effective but condensation should be considered.

(d) *Ground Floors*
    (i) *Suspended timber floors.* The temperature differential, and thus the

* See Section 3.7 ('Thermal Mass').

Table 3.7.   *Thermal conductivity of masonry materials* [40]

Bulk dry density kg/m^3	Thermal conductivity W/m °C		
	Brickwork protected from rain: 1%^{a}	Concrete protected from rain: 3%^{a}	Brickwork or concrete exposed to rain: 5%^{a}
200	0·09	0·11	0·12
400	0·12	0·15	0·16
600	0·15	0·19	0·20
800	0·19	0·23	0·26
1 000	0·24	0·30	0·33
1 200	0·31	0·38	0·42
1 400	0·42	0·51	0·57
1 600	0·54	0·66	0·73
1 800	0·71	0·87	0·96
2 000	0·92	1·13	1·24
2 200	1·18	1·45	1·60
2 400	1·49	1·83	2·00

^{a} Moisture content expressed as a percentage by volume.

heat losses through the floor, will vary depending on the use of the space below the suspended floor. An openly vented space to the outside will produce the same temperature difference as elsewhere on the weather skin. If the space is enclosed but unheated its temperature can be calculated using a resistance split method [41]. In Europe the renowned wine cellars at a constant year-round temperature of 10 °C (50 °F) give uniform heat losses. In the United States, where furnaces are usually placed in the cellar, the same constant cellar temperatures are applicable. Insulation instalment is usually complicated by the network of water and heat distributing pipes and ducts. Waterproof insulations, such as expanded polystyrene, are usually ruled out by fire regulations. Reflective shield insulation, such as that developed in Holland [42], although highly effective, if placed with reflective side facing up will soon become covered with dust and no longer reflect infrared radiation. A multiple-layered reflective insulation will perform well and so will polythene (polyethylene) covered fibreglass insulation. A low-emissivity surface facing the cellar will improve the thermal performance of the suspended floor.

ii) *Solid floor*. Site investigation will establish the level of the water table. If the groundwater table is high, heat will be rapidly stripped away from the

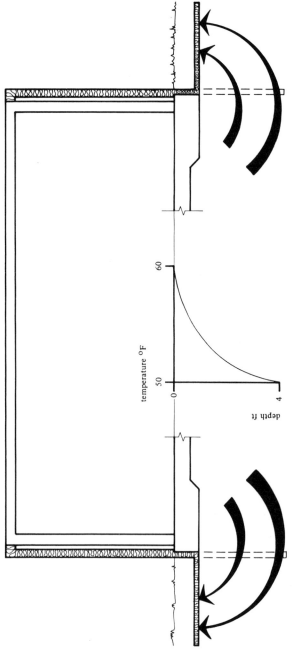

Fig. 3.40. Edge insulation and heat flow into the ground.

building through the floor. In such conditions there is no alternative but to insulate below the entire floor.

The groundwater temperature determines the temperature differential across which heat is lost. The groundwater temperature for each month can be measured from the cold water tap (provided that the water comes from the mains).

Where a high water table is not present, the ground below the building may be incorporated into the thermal mass by insulating against heat losses at the edges. Effective insulation of the edges of a building in such a location will result in negligible heat losses through the solid floor.

In examining heat movement between two layers of any material(s), the quantity of heat being conducted is proportional to the temperature difference between the two layers. If the ground beneath the building is divided into many layers, the resulting temperature differentials between the layers, and thus heat flow downwards, diminish exponentially (Fig. 3.40) until, at a depth of 1·2 m (4 ft), there is negligible heat flow. In view of this, the edge losses become the only problem. Figure 3.40 shows that these can be contained by insulating either horizontally or vertically, provided that the same paths of heat flow are intercepted.

3.13. INFILTRATION

The Hittman Report [34] concludes that infiltration accounts for some 28 % of the total seasonal heat losses in the majority of houses built in America today. A very well-insulated house in America may have infiltration losses of approximately 38 % of the seasonal heat losses. Figure 3.41 shows the changing infiltration rates for a standard and then a well-insulated and weather-stripped dwelling in England and America.

Thus, by substantially reducing the infiltration rate, 90 + % solar heating is easily attainable. The problem is that infiltration is reduced to this extent with great difficulty. The $1\frac{1}{2}$ air changes per hour for standard construction in America are reduced to about $\frac{1}{2}$ air change per hour by means of good weather stripping around openings and by caulking construction cracks.

Such a rule-of-thumb method is traditionally adopted because, until recently, accuracy in assessment of this air change loss has not been important. The alternative to this method is the crack method, which determines the heat loss by knowing the infiltration coefficient of the gap, the length of the gap, the degree of exposure of the building and the wind

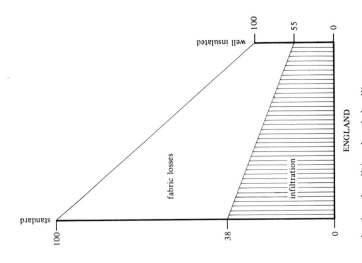

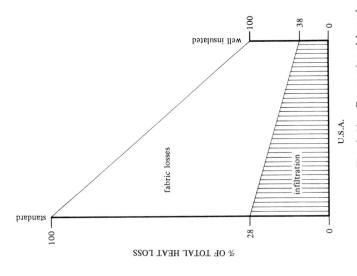

Fig. 3.41. Proportional heat losses in standard and well-insulated dwellings.

peed. Both methods call for a sixth sense of what is happening, largely based on comparison with measured test data. Because infiltration carries uch weight in a passively heated building, it is necessary to take a more detailed look at the methods both of calculating the heat losses to nfiltration and of determining the influencing factors involved.

The Crack Method

Where new windows are being installed, provided that relevant manufacturer's data are available, the infiltration air flow can be calculated rom the following relationship [27]:

$$V_a = L \times C \times (\Delta p)^{0.63} \qquad (24)$$

where V_a = volume of air flow, m^3/h (ft^3/h),

L = gap length, m (ft),

C = infiltration coefficient = volumetric air flow per metre (foot) of opening joint per unit pressure difference, $m^3/h\,m\,N/m^2$ ($ft^3/h\,ft\,lbf/ft^2$),

Δp = pressure difference across the window.

The effect of wind on a building is to create a differential pressure on the windward face of 0·5–0·8 times the velocity pressure of free wind. This is reinforced by a suction pressure of 0·3–0·4 times the velocity pressure on the leeward face. The combined pressure difference across the building is therefore approximately equal to the wind velocity pressure. In eqn (24) Δp refers to half of this, being only the pressure difference across the window; and the crack length taken is for all windows of the building. British Meteorological data are based on readings in open country at a height of 10 m (30 ft). Local velocities will be higher than this for windows at a greater height and lower for lower windows or for buildings sited in built-up areas. Mean monthly wind speeds in England show an average of approximately 5 m/s (11 mile/h) in winter [43] which will be approximately 3 m/s (7 mile/h) in surburban areas. This represents wind pressures of 45 to 15 N/m² (0·3–0·1 lbf/ft²) respectively.

Tests by the UK Electricity Council Research Centre (ECRC) on unoccupied houses in England, using a tracer gas [44], have shown that, although infiltration rates are closely linked to wind velocity in a normal house, in the well weather-stripped house ventilation rates are mainly dependent on internal to external temperature difference. This leaves the crack method somewhat wanting as a design tool. Moreover, neither calculation method includes infiltration due to the behaviour patterns of

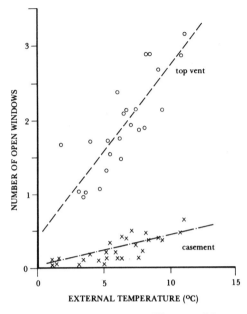

Fig. 3.42. Number of windows open at different outdoor temperatures [45].

the inhabitants. Window and door opening can, particularly in housing contribute a significant additional seasonal heat loss.

Dick and Thomas in 1951, in a thorough but neglected study [45] revealed the British tendency to open the windows in mild weather. The number of windows open increased linearly with outdoor temperature, with the numbers slightly reduced in high wind conditions (Fig. 3.42). The effec was to increase the air change rate per hour from approximately $1\frac{1}{2}$ at 0 °C ($32\,°F$) to $3\frac{1}{2}$ at 12 °C (54 °F) (Fig. 3.43). Such behaviour patterns can largel be explained by the high humidity of the air during the heating season in England. For most of the English winter the outdoor air is approximatel 90–100 % saturated with water vapour [46]. Brundrett [27] illustrates how the infiltration rate increases because of the behaviour of the inhabitants in the ECRC Bromley field trials (Fig. 3.44). The solid straight line on the graph in Fig. 3.44 shows where an additional $\frac{3}{4}$ of an air change per hour wil coincide with the measured behavioural infiltration loss pattern. The theoretical consumption with closed windows was computed using $1\frac{1}{2}$ ai changes per hour as the infiltration rate. Thus in normal buildin construction we can probably expect something like $2\frac{1}{4}$ air changes per hou over the heating season in England. This additional infiltration rate may

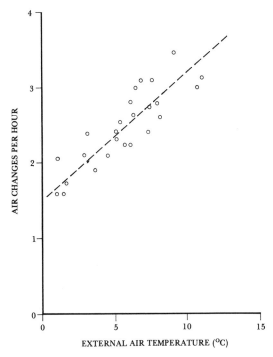

Fig. 3.43.   Air change rate related to open windows [45].

change with increased insulation, weather stripping, and increased use of man-made fibres for furnishings. Single glazing and natural fibres will often act as dehumidifiers in an English home. If double glazing or transparent insulation are used, condensation will not occur and more ventilation may be required to reduce a build-up of humidity levels in the air. Brundrett [47] argues that the increased use of man-made fibres will also increase the user infiltration rate. The reasoning is that natural fibres within a building tame the daily humidity fluctuations because they absorb far more moisture than do man-made fibres. There are no figures available to verify these probable changes in infiltration rate nor to suggest the magnitude of the change involved. Therefore, until this is known, it is suggested that the same increment of $\frac{3}{4}$ of an air change per hour be used for that portion of the infiltration caused only by behavioural patterns of the occupants of a well-insulated house in England.

The ECRC tracer gas infiltration tests [44] have shown that the air change rate for a well weather-stripped unoccupied house is on average 0·18 air

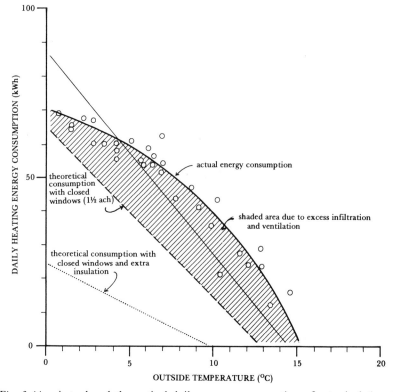

Fig. 3.44. Actual and theoretical daily energy consumption of a typical three bedroomed row house in England [27].

changes per hour. Thus a figure of 0·93 (0·18 + 0·75) air changes per hou can be used for an occupied, well-insulated and weather-stripped house i England. Tracer gas tests on the MIT Solar 5 building revealed a infiltration rate of 0·5 air changes per hour [48]. This suggests that this is a area of knowledge in need of further research and definition.

The obvious progression from here is to determine how much air i actually needed by the inhabitants for comfort and life support, and to see i there is any difference between this and the 0·93 air changes per hou derived above for the weather-stripped house (0·5 in the USA). It i necessary to look at each of the constituent needs supplied by fresh air

(*a*) *Oxygen.* A wide range of percent oxygen content in the air can b tolerated. At sea level the percent oxygen in the air can be reduced from

21% to 13% without any noticeable effect on the rate of breathing. Reduction below 12% will cause an increase in breathing rates. However, this is unlikely to be encountered, because we are far more sensitive to the concentration of carbon dioxide in the air.

(*b*) *Carbon dioxide.* Outdoor air contains 0·03% of carbon dioxide. When a concentration of 2% carbon dioxide is reached in inhaled air the depth of breathing increases. At concentrations of 3–5% there is a conscious need for increased respiratory effort and the air becomes objectionable. Concentrations of over 6% carbon dioxide in air are dangerous [49]. In England the maximum recommended concentration of carbon dioxide for habitation is 0·5% [43]. This gives a generous safety margin for breathing comfort. The resulting fresh air need for comfort from this parameter is 3·8 m^3/h per person. For housing this is approximately equivalent to $\frac{1}{4}$ of an air change per hour.

(*c*) *Odour control.* Infiltrating air dilutes odours. Experiments conducted by Yaglou *et al.* in the USA [50] show that there is no sex difference in the emanation of odours if no perfume is worn, but that age is important since children below the age of 14 create more objectionable odours. Also, the strength of odour was shown to be strongly related to the time since the last bath.

Figure 3.45 shows that $\frac{1}{3}$ of an air change per hour is acceptable for both carbon dioxide and odour levels, providing that the inhabitants are non-smokers and that there is a space of 8 m^2 (85 ft^2) per person. Presumably this air change rate can be substantially reduced if it is possible to remove odours and carbon dioxide by some means other than the introduction of fresh air.

In England a different problem occurs at this point. If the air change rate is reduced to $\frac{1}{3}$ per hour, we are introducing just over 5 m^3/h of fresh air per person and, as shown in Tables 3.8 and 3.9, the resulting relative humidities will be uncomfortable for a large proportion of the heating season. Current comfort guides [43] recommend a range of 40–70% relative humidity. Consequently the inhabitants will open windows. This implies that a method of dehumidifying the air is required for an English home. It is thought that a solar dehumidifier on the lines of that suggested by Wellesley-Miller [51] could, with modification and testing, possibly do this job.

Such moisture problems do not occur in American winters. Therefore, theoretically, in the United States an air change rate of 0·3 per hour would

Table 3.8. The influence of air change on relative humidity for an internal temperature of 17°C [47]

Fresh air per person m³/h	kg/h	Extra moisture added by person g/kg air	External Temperatures −5°C Initial moisture g/kg air	−5°C Final g/kg air	−5°C R.H. at 17°C	0°C Initial moisture g/kg air	0°C Final g/kg air	0°C R.H. at 17°C	5°C Initial moisture g/kg air	5°C Final g/kg air	5°C R.H. at 17°C	10°C Initial moisture g/kg air	10°C Final g/kg air	10°C R.H. at 17°C	15°C Initial moisture g/kg air	15°C Final g/kg air	15°C R.H. at 17°C	20°C Initial moisture g/kg air	20°C Final g/kg air	20°C R.H. at 17°C
5	6·0	6·6	2·2	8·8	73%	3·3	9·9	82%	4·6	11·2	93%	6·3	12·9	sat.	8·3	14·9	sat.	9·2	15·8	sat.
10	12·1	3·31	2·2	5·5	46%	3·3	6·6	54%	4·6	7·9	65%	6·3	9·6	80%	8·3	11·6	96%	9·2	12·5	sat.
20	24·1	1·66	2·2	3·9	32%	3·3	5·0	41%	4·6	6·3	52%	6·3	8·0	66%	8·3	10·0	83%	9·2	10·9	91%
30	36·2	1·1	2·2	3·3	27%	3·3	4·3	36%	4·6	5·7	47%	6·3	7·4	61%	8·3	9·4	78%	9·2	10·2	86%
40	48·2	0·83	2·2	3·0	25%	3·3	4·1	34%	4·6	5·4	44%	6·3	7·1	59%	8·3	9·1	76%	9·2	10·0	83%

Assumptions: one sedentary person contributes 40 g/h moisture; specific volume of air = 0·83 m³/kg.

Table 3.9. The influence of air change on relative humidity for an internal temperature of 21°C [47]

Fresh air per person m³/h	kg/h	Extra moisture added by person g/kg air	External Temperatures −5°C Initial moisture g/kg air	−5°C Final g/kg air	−5°C R.H. at 21°C	0°C Initial moisture g/kg air	0°C Final g/kg air	0°C R.H. at 21°C	5°C Initial moisture g/kg air	5°C Final g/kg air	5°C R.H. at 21°C	10°C Initial moisture g/kg air	10°C Final g/kg air	10°C R.H. at 21°C	15°C Initial moisture g/kg air	15°C Final g/kg air	15°C R.H. at 21°C	20°C Initial moisture g/kg air	20°C Final g/kg air	20°C R.H. at 21°C
5	5·9	6·72	2·2	8·9	56%	3·3	10·0	64%	4·6	11·3	72%	6·3	13·0	83%	8·3	15·0	96%	9·2	15·9	sat.
10	11·9	3·36	2·2	5·56	35%	3·3	6·6	42%	4·6	8·0	51%	6·3	9·6	61%	8·3	11·6	74%	9·2	12·5	80%
20	23·8	1·68	2·2	3·88	24%	3·3	5·0	32%	4·6	6·3	40%	6·3	8·0	51%	8·3	10·0	64%	9·2	10·9	68%
30	35·7	1·12	2·2	3·32	21%	3·3	4·4	28%	4·6	5·7	36%	6·3	7·4	47%	8·3	9·4	60%	9·2	10·3	66%
40	47·6	0·84	2·2	3·04	19%	3·3	4·1	27%	4·6	5·4	35%	6·3	7·1	46%	8·3	9·1	58%	9·2	10·0	64%

Assumptions: one sedentary person contributes 40 g/h moisture; specific volume of air = 0·843 m³/kg.

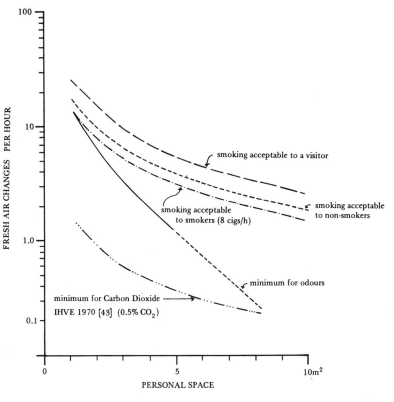

Fig. 3.45.   Fresh air changes as a function of personal space (ceiling height of 2·4 m assumed) [47].

e acceptable in the controlled environment described above. For alculation purposes it is assumed that the minimum comfortable air hange rate attainable in America is about half an air change per hour.

3.14.   OVERHEATING

s mentioned at the beginning of this chapter, the main purpose of passive olar heating design is to tame the daily and seasonal fluctuations in room ir temperature. If this is not achieved and the collection area is oversized, wo important side effects will result: first, the inhabitant will become ncomfortably hot on clear days, even in the heating season, and second, this will cause opening of windows to alleviate the discomfort. Such opening of windows within the heating season means that some of the potentially collectable solar energy is wasted; it is not stored for later use. Thus being able to test a design for overheating will have implications concerning both comfort and the overall annual heat loss/gain balance. If the building will not overheat during the heating season, simple heating load analyses may be conducted with reasonable accuracy. If the building will overheat, the hourly annual heating load simulation will be necessary to determine when the space overheats and how much of the potentially collectable energy is wasted when venting occurs.

Thus the behaviour of the room air temperature is possibly the most important piece of information available to the designer of a passive solar building.

There are many different analytical methods available for providing the designer with this information. Only two such methods are provided within this book but each of the methods known to the author will be discussed here to give an overall picture of the field.

(a) *Large Computer Programs*

The finite difference (and element) thermal network methods are used for any serious research work in the passive solar field. Both consist of defining any passive system design in terms of nodes. Each node has associated with it resistances in all directions or modes of heat flow, and a thermal capacitance. In a two-dimensional thermal network a massive element (e.g. a Trombe–Michel wall) is divided into many layers, each layer being represented by one node. In this way the dynamic flow of heat is modelled with accuracy. The information supplied by this method is the temperature at each node. Extreme flexibility and accuracy are the advantages of such method, although more accurate modelling requires a three-dimensional thermal network. The architecture of the nodes is not a predetermined constraint but the difficulty arises when deciding what to include in the model. The first example of this type of model in passive solar analysis was developed by Davies [52] in his analysis of the Wallasey School. Then came 'Pasole' by Balcomb and McFarland at Los Alamos Scientific Laboratory [53] and 'Thermal' by Johnson at MIT [16]: 'Pasole' was developed specifically for problems where the three-dimensional movement of the sun around the building being modelled was of no concern. This is why the Trombe–Michel wall and the water wall were the first passive types analysed at LASL. 'Thermal' was developed to model a direct gain situation where movement of the sun through the room is of prime significance. Both rograms require large computers and are therefore of little direct use to the esigner. These programs have been reduced to forms which can be used by esigners and these will be described under heading (b). LASL is at present eveloping another large program called 'Sunspot Level II' which will ynamically model direct gain spaces. The outcome of this work and its ssociated design methods are awaited with interest.

b) Small Programs for Programmable Calculators

(i) *Teanet* [54]. This was developed by Kohler of Total Environmental Action (New Hampshire, USA) on the same principles as those used in Pasole', and has been validated [55]. This program is for use on the TI–59 Texas Instruments) programmable calculator plus printer. It will model a maximum of seven nodes and when used on a calculator is only practical for modelling one or two days (rather than an entire year) because each day akes approximately 30 minutes to model. Flexibility is again the great dvantage of this method, for the seven nodes may describe most passive ystems. The systems not describable by this program are any system which elies on air movement to store or redistribute heat (e.g. attached sunspace ising a temperature differential controlled fan to pull energy into the ouse), and non-diffusing direct gain systems where movement of the sun hrough the space is of importance. An additional disadvantage of this nethod is that a number of days has to be modelled before equilibrium is eached, otherwise erroneous results occur. It is difficult to use this program or modelling more than one day because the hour numbers modelled here re from 0 to 23. Thus the first hour of the next day (again zero) is hour 23 of he two-day period, because the program prints out the start values and all esults on the second day are therefore one hour out of synchronization. It vas also found, in the validation procedure [55], that 'Teanet' produced esults which were consistently shifted by one hour even on the first day nodelled. This is due to the sequencing used within this program. The PSP rogram was written as an outcome of this finding.

(ii) *PSP—passive simulation program.* This is a thermal network rogram written by the author for use on the TI–59 programmable alculator plus printer. It can model a maximum of seven nodes and is ased on the mathematical principles outlined by Kohler et al. [54]. This rogram is presented in Appendix B.

(iii) '*Solarcon*'.* This, like 'Teanet' and PSP, is a finite difference thermal

Graeff, Solarcon Program ST33, 607 Church, Ann Arbor, MI 48104, USA.

network program developed by Graeff for use on the TI–59 programmable calculator plus printer. The program has not yet, to the knowledge of the author, been externally validated.

(*iv*) '*Pegfloat*', '*Pegfix*' [56]. Developed by Glennie of the Princeton Energy Group for both the TI–59 and a Hewlett–Packard 67/97, these programs are a simplification of the thermal network concept using three nodes to model the passive system. These programs have been validated [55]. It is thought that using only three nodes will give misleading results when applied to non-diffusing direct gain systems where distribution of thermal mass is of importance. This is the same criticism as that levelled against the Admittance Procedure below. 'Pegfloat'/'Pegfix' can be used to model only direct gain systems.

(*v*) *PDP1—passive design program* [19, 20]. PDP is a program for use on the TI–59 which was developed by the author from 'Thermal' (see heading (a)) and is specifically for modelling a south-facing direct gain system. This method has been validated [55] and is presented in Appendix C. The detailed calculations involved in this method are reported elsewhere [19, 20]. The version of this program presented in Appendix C, although mathematically identical to the previously published version, is slightly easier to use.

(*c*) *The Admittance Procedure* [57]
This is a manual method of determining daily room air temperature fluctuations developed by the Building Research Establishment (UK) to design against summertime overheating. It is a thoroughly researched method in which the user describes the materials of the room concerned in terms of thermal admittance (similar in concept to U–values). The thermal admittance of a material makes allowances for the dynamic response of these materials to heat flow. A major problem with this method is that all thermal masses within the space are lumped, and therefore it will not respond to the changes in room air temperature fluctuation caused by sizing of thermal masses, as shown in Fig. 3.16 (Section 3.7, 'Thermal Mass'). The method is appropriate for diffusing direct gain systems. It is hoped that some further research will be conducted to advance this method so that it may be used for all types of direct gain systems.

The programs listed in Appendices B and C are to be used to place an upper limit on the open-ended design method for system sizing (presented in Section 3.17 ('System Sizing and Auxiliary Heating').

3.15.   SHADING DEVICES

The visual alternative and technical aspects of shading devices are fully described elsewhere [58]; thus little time will be devoted to this aspect of solar control here.

Fixed shading devices have some inherent disadvantages, the first of which is due to their inability to allow for the lack of synchronization between the heating season and the altitude of the sun. If a fixed shading device is designed for Boston to prevent overheating in September, the result is that solar energy will be rejected to the same extent in March. March, unlike September, has a relatively high demand for solar heating and such control would be unwelcome. Fixed shading devices will also usually cast a shadow on a portion, albeit a small portion, of the collection window for most of the heating season. This is expensive in terms of cost of glass and higher energy losses than gains over this area of shaded window.

The alternatives to fixed shading devices are either movable devices or the use of deciduous vegetation. Both of these have possible economic constraints. Unless the vegetation is placed on a trellis above the collection window, there is a reduction in the quantity of solar energy received even after defoliation. This will necessitate the expense of additional collection windows. As for movable shading devices, these will be usually more expensive than fixed devices.

Large east- or west-facing windows will require shading devices to prevent excessive heat gains in the spring, summer and fall. Due to the way in which the sun appears to move across the sky, shading devices for these windows will be closely spaced, vertical and, in plan, at an angle to the window.

South-facing windows will usually need only horizontal shading devices.

3.16.   SYSTEM EFFICIENCY

The efficiency of a passively heated building is in some ways comparable to that of a flat plate solar collector. However, there are certain prominent differences. The following is a discussion of the similarities and differences and a method for determining the efficiency of a passive heating system.

When considering the efficiency of a flat plate collector, the heat losses of the building it heats are not included in the calculation. Similarly it is proposed that the losses involved in the efficiency calculations of a passive

system be only those through the collection window. More difficult t
handle, however, are the internal gains, the solar gains from other window
and the infiltration losses of the building. These could be apportioned to
unit area of total weather skin and thus included in the efficienc
calculations as negative heat losses through the collection windows; but it i
suggested that they be ignored for the sake of simplicity.

The active system is thermostatically controlled to cut off or drain dow
at the end of the day or when the sun goes behind the clouds. Thus the fla
plate collector is operating approximately 8 out of the 24 hours in a da
whereas, in terms of losses, the passive solar system operates for all 2
hours. In view of this, one can expect the efficiency of a passive system usin
normal building materials to be nominally $\frac{1}{3}$ of the efficiency of a flat plat
collector. Certain factors counteract this reduction in efficiency. Th
passive building is operating at lower temperatures than a flat plat
collector; this reduces the heat losses from the system by approximately
factor of 2. There are the additional benefits of the passive system, a
mentioned in Chapter 1, such as the longer hours of collection and th
collection of diffuse energy on an overcast day. Thus, considered in broa
terms, the passive system probably has an average efficiency of 85 % of th
average efficiency of an active system built of similar conventional materia
and with the same orientation.

The efficiency of the passive collection system is determined by th
following equation:

$$\eta = (Q_g - (Q_{1h} \times 24))/I \qquad (25$$

where η = the efficiency of the collection system,
　　　Q_g = the net solar energy gain through the collection window on
　　　　　clear day = $(I \times \text{transmission}) - \text{back-reflected losses}$,
　　Q_{1h} = the heat loss per hour through the collection window
　　　　　= $U\text{-value} \times \text{area} \times \Delta T$,
　　　I = the total clear day incident solar energy on the collectio
　　　　　window surface.

In an active system the quantity of solar energy collected is reduced by th
absorptance of the collector surface. In a passive system a comparabl
reduction is due to reflected back losses. At the moment the estimation o
these back losses is very much a black art and is further discussed in Sectio
3.8 ('Calculating Solar Gains').

When movable insulation is used in a passive system, the value of Q_{1h} i qn (25) will differ for the two operation modes of the system. Thus changes to this value must be made when calculating the efficiency of such a system.

3.17. SYSTEM SIZING AND AUXILIARY HEATING

There are many design considerations which go into the determination of the area of collection window used in a system. Hypothetically, if there are no such design constraints, the aim would be to achieve 90–100% solar heating based on heating load calculations. Such calculations are discussed in Section 3.11 ('Heating Load Calculations') and further in Section 3.18 ('Cost Analysis'). The consequences of the system sizing are then tested by means of one of the programs described in Section 3.14 ('Overheating') to examine whether or not the air temperatures produced fall within the comfort zone [59–61].

A solar heating fraction of 100% is not usually possible, unless seasonal storage is built into the system, because of the undependable weather patterns. Overcast weather may last for several days; the longest period of overcast weather recorded in Boston was 29 days and in London 30 days. Such spells can never be accounted for by a solar heating system without making the system grossly uneconomical. The highest average solar heating fraction one can hope successfully to achieve is probably 97%. Because of this a solar heated system usually has to bear the brunt of additional capital expenditure on a back-up heating system.

Such an auxiliary heating system will vary in capital expenditure from full central heating, when the solar heating fraction is low, to electric, gas or wood-burning heaters when the solar heating fraction is high. The large running costs of electric heating will not be a burden if the solar heating fraction is high. In the case of rehabilitation, reuse of existing heating systems may enhance the economic feasibility of utilizing solar energy for space heating.

Electric storage heating using off-peak electricity is not suitable for use with passive solar heating. This is because the two systems will fight each other and inhabitants will be forced to open windows on sunny days. It may be possible to overcome this problem by having the switching mechanism for the storage heaters tied into a weather forecast: when a sunny day is forecast the heaters will be switched off.

The auxiliary heating system for a passive solar house is sized in the conventional way. It should have an output which will meet winter design conditions without solar gains.

3.18. COST ANALYSIS

The costing aspect of a solar heating system is probably the single most important aspect within the entire field of solar heating, because upon th relies the success or failure of the system. Thus in order to provide a useful design tool for cost analysis, reliable heat load and cost benefit calculatio methods must be sought.

The basis of the heating load calculations used in the design metho presented in this handbook is developed in Section 3.11 ('Heating Loa Calculations'). The cost benefit calculations are based on well-establishe methods of Life-cycle Costing and are developed elsewhere [62, 63]. Th use of life-cycle costing methods makes it possible to compare the capit; expenditure of a solar heating system with the money saved by it each yea due to a saving in fuel use. Corrections are applied to the money saved eac year to account for the effects of inflation and for the rising price of foss fuel. This method gives an estimation of the system's worth in terms of th present value of the money saved.

These calculations are time-consuming and thus present an obstacle i the way of their use by building designers. It was thought that by puttin such a design method on to a TI–59 programmable desk-top calculator, thi obstacle would be removed. The resulting program and the procedure fo its use are presented in Appendix D.

The following methods are employed in this cost benefit program.

An approximation of the heating load per month is computed followin the steps described in Section 3.11 (eqns (19)–(23)). This is a approximation only, because it uses mean monthly temperatures t compute the number of degree days.

Equation (19) is difficult to solve because it contains two interdependen variables: the degree day base temperature (balance point) and the interna heat gains per day. Because the value of the internal heat gains per da varies each month (due to the sun's altitude, the percent sunshine, an inhabitant behaviour patterns), determining which months are appropriat for the calculation is dependent on there being a heat loss during that montl which is in excess of the internal gains. This is thus dependent on the degre day base temperature. But the degree day base temperature is in turi dependent on the internal gains per day figure. Therefore iteration of eq; (19) is necessary to narrow down on the correct result. The program tests t see if the heat gains are less than the heat losses each month. Only unde these circumstances do the heat gains beome included in the heat gains pe day value in eqn (19).

Table 3.10. *Balance point temperatures with corresponding degree day figures for a typical winter in London, after Brundrett [27]*

Balance point temp. (°C)	Degree days	Difference
−5	0	0
−4	1	2
−3	3	5
−2	8	8
−1	16	13
0	29	22
1	51	32
2	83	45
3	128	60
4	188	78
5	266	98
6	364	118
7	482	139
8	621	161
9	782	178
10	960	194
11	1 154	207
12	1 361	218
13	1 579	224
14	1 803	229
15	2 032	232
16	2 264	233
17	2 497	—

If a passively heated building with a high solar fraction is being modelled, it could happen that in none of the heating months do heat losses rise above the heat gains figure. In such a case the program takes an average of the heat gains per day during December and January, and uses this value to continue computation.

Once the degree day base temperature is displayed on the calculator, the corresponding number of degree days is entered into it and the program is set in motion again to calculate the resulting heating load per year. The method for developing a degree day table is described in Section 3.11. An example of such figures (for London) is shown in Table 3.10. It is useful, as shown here, to keep a note of the difference between the degree days for one balance point temperature and those for the next one up. This is used when, as is often the case, the degree day base temperature (balance point) displayed is between two whole number temperature values.

Having computed the heating loads and costs with and without improvements, the cost benefit of the improvement is calculated by the program. The method used here for this is based on standard methods of discounted cash flow (DCF) analysis. The cost benefit is the present value of the energy saved divided by the capital cost of the improvement. The present value of the energy is computed as the sum of a geometric series of discounted savings (eqn (26)). If the net present value of the total energy saved over N years due to an improvement is greater than the initial cost of implementing the improvement, then the improvement is cost effective. In other words, if the cost benefit is greater than 1 the improvement will be cost effective, and an approximate payback time can be found by dividing N (eqn (26)) by this cost benefit figure.

$$\mathrm{NPV} = F \times \frac{a(a^{N_y} - 1)}{a - 1} \tag{26}$$

where NPV = the net present value of the money saved over N_y years,

F = current year's value of the annual fuel savings,

$a = (1 + g)/(1 + r)$,

g = the 'real' rate of increase of fuel prices (expressed as a fraction),

r = the 'real' discount rate (expressed as a fraction),

N_y = either the lifetime of the improvement or the number of years of borrowing, whichever is the shorter.

To choose appropriate values for the variables in eqn (26) is a considerable problem because it involves a certain amount of forecasting. The following is a discussion on the values to be used for some of these variables [63].

Maintenance costs should be subtracted from the value of the annual fuel saving.

Choosing a figure for the discount rate is difficult. If the scheme is carried out by an individual using his own capital, the discount rate should represent the rate of interest (in real terms) gained by investing the money elsewhere. Depending on the country's inflation rate and tax system on unearned income, this interest rate can be lower than expected. In England in recent years, the interest rate (in real terms) has been as low as 0–2 % and often negative. The 'real terms' refer to the percentage above the inflation rate percentage; it also takes account of the effects of taxation. Often the opportunity value of this capital is much more valuable than its interest returns. This opportunity value will disappear if the money is invested in, say, insulation.

For individuals without private funds, the discount rate will be that of the interest rate (in real terms) on borrowed money. An examination of the past trends in home mortgage rates in England has shown that they are about 2% greater than the rate of inflation from 1955 to 1970 (since 1973 the mortgage rate has been less than the rate of inflation). Thus it seems very worthwhile to make energy-saving improvements which give returns greater than 2% of the capital expenditure.

For public sector work in England the Treasury has established a test discount rate (TDR) which it applies to its investments. This is currently 5% p.a.

Going by the figures mentioned above, for the individual investor in England with funds of his own, a discount rate of 3% would be more realistic than the 5% p.a. used by the Treasury.

The second difficult figure to establish is the real rate of increase in the price of fuel. From the figures of a wide range of demand and supply scenarios, it would appear that the real price of fuel will slightly more than double by the year 2000. If this rise occurs smoothly, then a 4% p.a. real rate of increase in fuel prices can be expected. This is the increase above the rate of inflation.

The value N_y in eqn (26) will represent the shorter of either the life expectancy of the improvement or the number of years over which the mortgage extends.

If, during execution of the program, the cost benefit displayed is less than 1 and it is attempted to compute the payback period, the display will read 1 000 000 years, indicating that the improvement is not economically viable.

When the discount rate and the fuel cost spiral rate are equal, there is a straight line payback relationship. Thus in the program, under such circumstances, the payback period will be displayed in place of the cost benefit.

Because of the uncertainties of the variables used in eqn (26), this section of the program has been designed so that it is possible to compute only this program routine. Thus different values for the fuel spiral rate, the discount rate, the number of years or the area and cost of improvement may be inserted and the cost benefit recalculated.

3.19.   GLARE

Glare is one of the major design conflicts inherent in both the use of natural lighting and of direct gain passive solar energy collection. It is therefore important to be able to understand its nature and identify its causes and remedies.

Glare is related to the contrast in one's visual field, and can be avoided by simply turning away from the light source. The most painful glare is when the sun shines directly into the eyes. Reflection off water causes this type of glare. Upward reflection will cause more discomfort because the eyes are unshaded by the eyebrows. This is the easiest type of glare to deal with conceptually, because if it exists it will be intolerable. Glare is also the result of effects more difficult to define or counteract.

Disability or veiling glare is detected as the area of indistinct vision around a bright light source. This is caused, as the name suggests, by a scattering of the light in the eye fluids, which casts a veil of light over the image formed on the retina. A simple experiment will illustrate this effect. Look at the mullions of a window—it will be noticed that much of the detail of them is indistinct. Now shield the view of the window and visual discrimination should improve. Disability glare is additive, i.e. the brightness of the veil produced by 10 identical glare sources is 10 times the brightness of the veil due to one such source on its own. Old people are particularly susceptible to disability glare because their eye fluids become cloudier with age; this causes more scattering and thus a brighter veil. Disability glare is seldom of great significance in architectural design other than serving as a pointer to errors in lighting design, such as placing a blackboard on the same wall as the classroom windows. In extreme cases disability glare becomes discomfort glare.

Discomfort or contrast glare is often experienced in conjunction with disability glare but they are separated in order to understand and quantify each. Discomfort glare, as the name suggests, causes discomfort rather than reduction in visibility. It is thought to be caused by a saturation of neural response in the retina due to large quantities of light. Because of the nature of this type of glare, there are serious architectural implications and consequently it is important to be able to assess the possibility of its occurrence. The luminance of large areas of sky visible through a window is one cause of discomfort glare. Another cause is excessive contrast in the visual field. If no artificial lighting is used, the interior levels of illumination will be lower in proportion to the sky brightness on a dull day and thus there will be more disability glare than on a bright day.

The glare constant and glare index are numbers given to quantify the glare caused by a light source. The glare index is a more recognizable and less cumbersome form of expressing the glare constant. The two are related by the following expression:

$$\text{Glare Index} = 10 \times \log_{10} \text{Glare Constant} \tag{27}$$

Until fairly recently, the methods of determining the glare constant of small light sources, developed by the Building Research Establishment [64], were used to find the glare constant for large windows. It has been discovered that a change in the size of a window may have no effect at all on the resultant glare. This occurs when the window is the only means of lighting in the room and is because with the change in area of sky seen through the window comes a proportional change in surround illumination. Thus a reduction in the area of the window will cause lower interior surface luminances and the contrast between this and the sky seen through the window is increased. Under certain conditions this will result in greater discomfort than that experienced with a larger window. Similarly any change in the transmission of the transparent membrane used will have no bearing on the glare constant. This means that, although commonly interpreted otherwise, any change in the transmission of the glazing material in a design will have no effect on the glare index of the designed space if both the surround illuminance and the luminance of the glare source change in equal proportions (i.e. there is no artificial lighting or side window with different transmission). Further numerical information may be found in references [65] and [66].

The following are a few guidelines to help avoid glare problems in a daylit room. If the room has windows on one side only, its depth should not exceed 2 to $2\frac{1}{2}$ times the height from the top of the window to the floor; and whenever possible additional daylight should be brought to the rear of the room to increase the average illumination level there and so reduce contrast. This additional lighting may be provided by either clerestory or side windows. Dark-coloured window bars, frames or window walls will all increase the contrast between them and the sky and hence increase discomfort glare. Large light-coloured splayed reveals to windows will reduce contrast glare.

Other effective methods of reducing discomfort glare are by the use of blinds, by having light-coloured floors and window walls, and by directing some additional lighting on to the main window wall.

A useful design tool for studying the effects of daylighting is a scaled model. The model should be large enough to enable one to put one's head inside it. Because the effects of daylighting scale without distortion, a model will give a good grasp of the probable feel of the designed space at different times of the day, and will reveal any problems of glare. A model may also be built with enough flexibility to enable comparative modelling of lighting design alterations. This method will be far more expressive than the use of calculations or drawings.

3.20. MONITORING

Passive solar architecture is in its infancy and because of this it is of great importance for passive solar designers to get feedback from their constructed and inhabited designs. Monitoring will provide information to check the accuracy of prediction methods being used for annual heating load analysis and for overheating assessments. It is also the only way in which the designer will learn from his design mistakes. Monitoring has a second purpose of being used to validate computer simulation models. These two functions of monitoring must be separated to identify the particular needs of each and to establish whether or not the means of collecting the data will be different for the two cases. The data requirements and proposed methods of collecting such data will be discussed in detail within this section.

(a) Monitoring for Computer Model Validation

For this type of work there is no need for the passive system to be lived in. This is a purely scientific operation and the presence of inhabitants will upset all readings by introducing an unquantifiable element. It is suggested that the testing systems should be more architecturally appropriate than the 1·6 m (5 ft) wide, 2·3 m (7 ft) deep and 3 m (9 ft) high test cells being used at the Los Alamos Scientific Laboratory in the USA. These dimensions would be more appropriate if increased to those of a small house, whose morphology is chosen to be typical of the local building stock.

The data required for validation work are as follows:

(i) hourly insolation incident on the glazing surface—total and diffuse. For more accurate modelling 15-minute totals of solar energy may be required,

(ii) hourly outdoor temperatures (shielded from solar radiation),

(iii) hourly auxiliary energy input,

(iv) up to 20 temperature sensors giving hourly readings. This number will increase depending on the size and complexity of the system,

(v) number of air changes per hour due to infiltration (taken as a one-off test(s)).

Typical examples of equipment required for the above data collection is listed below:

Example 1   Prices in UK (£)
Solarimeter by Kipp & Zonen for total solar.*   342·00
Solarimeter with adjustable shadow ring attachment*
for diffuse solar.   463·00

rinting integrators* for solar data (or in place of
these: 2 chart recorders* at £573·00 each).   1744·00 × 2

)ata logger 78101 Stereatronics by Norman
Saunders†
+ 33 sensors and connectors   425·00   ($850)
(40 input channels: 8 digital and 30 to Analogue-to-
Digital converter, recording done every hour).

Cassette tape recorder.   20·00

t should be pointed out that the decoding of the binary information
ecorded by the Saunders data logger is done by Saunders and presented on
paper at a nominal charge.

	Prices in UK (£)	
Example 2		
The same 2 solarimeters.	805·00	⎫
Microdata model M1600L data logger.‡	2955·00	⎪
2 integrating cards C1504 (for solarimeter data) at £115·00 each.	230·00	⎪
1 block of 10 cards consisting of:		⎪
1 thermister amplifier,	85·00	⎪
9 dual thermister multiplex cards (double scanning will allow 20 thermister inputs) at £40·00 each.	360·00	⎬ £4080
1 internal replay card for checking correct functioning.	240·00	⎪
Real time cards:		⎪
1 day card C1625,	75·00	⎪
1 h/min card C1345.	75·00	⎪
1 counter card if pulse counting (for monitoring auxiliary energy supply).	60·00	⎭

(This leaves 4 free card positions one of which can be
used for a header data card to label the tape (£75, or
£85 for one which occupies 3 slots)).

* From Enraf-Nonius Ltd, Highview House, 165–7 Station Road, Edgware, Middx
HA8 7JU, UK.
† From Norman B. Saunders, P.E., Experimental Manor, 15 Ellis Road, Sunshine
Circle, Weston MA 02193, USA.
‡ From Microdata Ltd, Monitor House, Station Road, Radlett, Herts WD7 8JX,
UK.

The Microdata model M1600L data logger will scan and select data for recording on to a $\frac{1}{4}$-in magnetic tape cartridge and will format the data so that it can be read directly by a FORTRAN program (ECMA/ANSI compatible with ASCII coding). The capacity of data storage on the cartridge is 66×10^3 readings per track. Each reading consists of four digits and a sign. There is a back-up battery provided—in case of power failure the logger can continue logging for the time taken to fill one track of tape. This data logger can be used as a peripheral to a computer for data replay and analysis—just plug it in!

Data collection by the use of strip chart recorders very soon becomes uneconomical if one includes the hours involved in extracting the data needed from the charts. However, the strip chart gives useful information about the time of a sudden change and the slope of this change. Such information is often lost by microprocessor based data loggers. The Energy Technology Support Unit (Harwell, UK) has recently conducted an exhaustive survey [67] of the data loggers available on the British market. As a result of this survey they are involved with the development of a data logger which will scan all channels continuously, record at set intervals, and also record when the scan notices a predetermined proportional change from the previous scan of the same channel. This in effect gives as much information as a strip chart recorder and the price being aimed at is £1500–£2000. The survey noticed also the same split in the market as in the two examples listed above. This division is caused by the expense of software and formatting. If the data is to be recorded in binary notation on a magnetic cassette the price of the data logger can be low (as example 1) but it will require decoding and reformatting for use in a data bank. This may be a logical way to approach the problem for it will require one decoding and reformatting base only which could be established by the Government. A consumer guide to data loggers is needed because of concern about the information omitted from the glossy trade literature.

The Epply pyronometer and the Lambda LI200S* ($95) pyronometer have been tested side-by-side at LASL and are giving identical results over the continuing test period [68]. This Epply pyronometer is approximately ten times more expensive than the Lambda LI200S.

Hourly data of auxiliary energy input may require the adaptation of a kWh, or other fuel source, meter to give a digital signal.

The air infiltration rate is determined experimentally by using a tracer

---

* From Li-Cor Lambda Instruments Corp., PO Box 4425, Lincoln, Nebraska 68504, USA. Price in 1979.

gas. Ten percent of the building air volume is replaced by helium. Helium may be used because its concentration can easily be compared with that of argon using a mass spectrograph, and it is not toxic. Because helium is lighter than air, fans are used to mix and distribute the helium (about one fan for every $10–15\,m^2$—$100–150\,ft^2$). Samples of room air are drawn into vacuum flasks every 30 min. A graph showing the decay time of the helium is then constructed. This will determine the number of air changes per hour due to infiltration [48].

(b) Monitoring for System Performance Feedback

The performance of a passively solar heated building cannot be viewed as an isolated scientific function. This is because passive solar design means a particular architectural approach and therefore the energy use behaviour pattern of the inhabitants is to a large extent dependent on the type of building they are in. Thus the monitoring of system performance has to be done with the inhabitants there. This complicates matters enormously and makes this type of monitoring potentially very expensive because patterns of appliance use and door and window opening will need monitoring to arrive at any valid conclusions. Such a monitoring system, if computer controlled, may well cost more than £30,000. An alternative to this approach is to have a statistically correct number of solar and non-solar houses so that inhabitant behavioural differences will be cancelled out. This approach is being followed at Pennylands, Milton Keynes, UK by the Milton Keynes Development Corp. and has resulted in sixty houses of each type. Each house will be in some way monitored; an expensive, and not necessarily accurate, approach to the problem.

The aim of work presently being undertaken in the United States [69] is to avoid these excessive expenses and still arrive at reasonably valid performance figures. The approach taken is to attempt to develop a data logger with the 10 input channels and full formatting for about $2000. The ten input channels would monitor the following data:

(i)   Solar radiation on a horizontal plane.
(ii)   Solar radiation on the collection plane.
(iii)   Shielded outdoor temperature.
(iv)   Indoor room temperature.
(v)   Auxiliary fuel energy consumption or fossil fuel burner time.
(vi)   Power to appliances.
(vii)   Domestic hot water heater.
(viii) Movable insulation (all open or all closed).

(ix) Venting of excess heat (forced vent).
(x) Operation of the fan to storage.

It is then suggested that a number of one time measurements will fill in most of the remaining unknown information. Hamilton [69] suggests the following measurements:

(i) Overall efficiency of heating system (not just combustion efficiency).
(ii) Consumption rate of pilot lights.
(iii) Calorific value of fuel used.
(iv) Heat transmission of the domestic hot water tank's jacket.
(v) Check the calibration of the thermostat.
(vi) Test the infiltration rate (this may be done several times over the winter).

Such strategies may well develop into a more acceptable method of collecting performance analysis data. It is thought that this type of research will yield results which will help to structure the problems within this facet of passive solar work.

REFERENCES

1. GEIGER, R., *The Climate Near the Ground*, Harvard University Press, Cambridge, MA, USA, 1959.
2. LOFTNESS, V. E., *Natural Forces and the Craft of Building: Site Reconnaissance*, MIT M.Arch. thesis, 1975.
3. BALCOMB, J. D., HEDSTROM, J. C., and ROGERS, B. T., 'Studying hot air: Design considerations of air-heating collector/rock bed storage solar heating systems', *Solar Age*, 1(2), 1976.
4. UK–ISES., *Solar Energy—a UK Assessment*, 1976.
5. AIA RESEARCH CORPORATION, *Solar Dwelling Design Concepts*, For the US Department of Housing and Urban Development, 1976.
6. ASHRAE, *Handbook of Fundamentals*, 1977.
7. IHVE, 'Guide A6', *Solar Data*, 1975.
8. ANDERSON, B., and RIORDAN, M., *The Solar Home Book*, Cheshire Books, USA, 1976.
9. HARRISON, D., 'Beadwalls', *Solar Energy*, **17**, 317–9, 1975.
10. JOHNSON, T. E., *et al.*, *Exploring Space Conditioning with Variable Membranes*, N.T.I.S., US Department of Commerce, Springfield, Va 22161, Report PB, 245137, 1975.
11. FAIRWEATHER, L., and SLIWA, J. A., *The Architects Journal Metric Handbook*, 3rd Edition, The Architectural Press, London, UK, 1969.
12. MORGAN, E. A., *Patent Specification 1022411, 29 March 1962*. Patent application No. 12351/61, 6 April 1961. Published 16 March 1966.

3. BALCOMB, J. D., *Designing Passive Solar Buildings to Reduce Temperature Swings*, LASL paper LA–UR–78–1316, Los Alamos, USA, 5 May 1978.
4. BALCOMB, J. D., HEDSTROM, J. C., and McFARLAND, R. D., 'Simulation as a design tool, *Passive Solar Heating and Cooling Conference Proceedings*, ERDA, LA–6637–C, Albuquerque, New Mexico, 18–19 May, 1976.
5. DAY, J., *Towards a Synthesis of Energy, Form and Use: New Forms of Solar Space Conditioning made Possible by the Use of New Materials*, MIT M.Arch. thesis, June 1977.
6. JOHNSON, T. E., 'Lightweight thermal storage for solar heated buildings', *Solar Energy*, **19**, 669, 1977.
7. ANDREJKO, D. A., *Energy and Office Architecture: A Synthesis of Emerging Passive Energy Technologies with Energy Efficient Planning and Design*, MIT M.Arch.A.S. thesis, 1977.
8. WRAY, W. O., BALCOMB, J. D., and McFARLAND, R. D., 'A semi-empirical method for estimating the performance of direct gain passive solar heated buildings', *3rd National Passive Solar Conf. Proc.*, San Jose, January 1979.
9. LEBENS, R. M., 'A design tool to assess room air temperatures in a passively heated space', *2nd National Passive Solar Conf. Proc.*, Philadelphia, PA, USA, 16–18 March 1978.
0. LEBENS, R. M., *Exploring Various Aspects of Passive Solar Energy Collection, with Particular Reference to the Rehabilitation of Nineteenth Century Row Housing in England*, MIT M.Arch.A.S. thesis, February 1978.
1. ANDREJKO, D. A., 'The Solar Modulator: a novel approach to direct solar gain in architectural applications', *New England Solar Energy Association Conf. Proc.*, Amherst, MA, USA, 1976.
2. LIBBEY–OWENS–FORD Company, *Sun Angle Calculator®*, Merchandizing Dept., 811 Madison Ave., Toledo, Ohio 43695, USA, 1974.
3. HITTMAN ASSOCIATES, *Multi-family Housing—Data Acquisition*, for the US Department of Housing and Urban Development, 1972.
4. SIVIOUR, J. B., *Design for Low Energy Houses*, Electricity Council Research Centre, UK, ECRC/M922, March 1976.
5. BILLINGTON, N. J., 'Thermal insulation and domestic fuel consumption', *BSE*, April, 23–4, 1972.
6. WOLF, R., 'Chauffage et conditionnement électrique des locaux', *Eyrolles*, 63, 1974.
7. BRUNDRETT, G. W., 'Some effects of thermal insulation on design', *Applied Energy*, **1**, 7, 1975.
8. HEAP, R. D., Data assembled at Electricity Council Research Centre, UK, from MR 415 and Electricity Supply Industry Statistics.
9. BASNETT, P., *Modelling the Effects of Weather, Heating and Occupancy on the Thermal Environment inside Houses*, Electricity Council Research Centre, UK, ECRC/M847, September 1975.
0. WHITESIDE, D., *Cavity Insulation of Walls: A Case Study*, British Building Research Establishment Current Paper 20/74, February 1974.
1. ROMIG, F., and LEACH, F., *Energy Conservation in UK Dwellings: Domestic Sector Survey and Insulation*, I.I.E.D. Energy Project, 10 Percy St., London W.1.

32. CORNISH, J. P., *The Effect of Thermal Insulation on Energy Consumption in Houses*, British Building Research Establishment, East Kilbride, UK, 1976.
33. Private communication with Prof. P. F. Chapman of the Open University Milton Keynes, UK.
34. HITTMAN ASSOCIATES, *Verification of the Time-Response Method or Heat Load Calculation*, for the US Department of Housing and Urban Development August 1973.
35. THOM, H. C. S., 'The relationship between heating degree days and temperature', *Monthly Weather Review*, **82**(1), January 1954.
36. THOM, H. C. S., 'Normal degree days below any base', *Monthly Weather Review*, **82**(5), May 1954.
37. BALCOMB, J. D., and MCFARLAND, R. D., 'A simple empirical method for estimating the performance of a passive solar heated building of the thermal storage wall type', *2nd National Passive Solar Conf. Proc.*, Philadelphia, PA 16–18 March 1978.
38. BUILDING RESEARCH ESTABLISHMENT, 'Condensation', *BRE Digests: Services and Environmental Engineering*, Digest 110, Cahners Publishing Co., Boston MA, USA, 1973.
39. IBID, 'An index of exposure to driving rain', Digest 127.
40. IBID, 'Standardised *U*–values', Digest 108.
41. BRYAN, A. D., *Performance Studies of a Working Solar Home*, The Black River Construction Co., Albany, Vermont, USA p. 25, 1977.
42. KOETSIER, J. E., *et al.*, 'Het KEMA—Onderzoek naar Energiebesparende Ruimteverwarmingssystemen (deel II), *Elektrochniek*, **54**, November 1976.
43. IHVE, *Guide 1970*.
44. SIVIOUR, J. B., and MOULD, A. E., 'A tracer gas method for the continuous monitoring of ventilation rates', *CIB S17: Meeting at Holzkirken*, Munich. 28–30 September 1977.
45. DICK, J. B., and THOMAS, D. A., 'Ventilation research in occupied houses', *J Institute of Heating and Ventilating Engineering*, **19**, 306–26, British Building Research Establishment. Crown Copyright. By permission of the Controller, HMSO, 1951.
46. HEAP, R. D., *Heating, Cooling and Weather in Britain*, Electrical Council Research Centre, ECRC/M631, June 1973.
47. BRUNDRETT, G. W., *Window Opening in Houses: An Estimate of the Research and Magnitude of the Energy Wasted*, Electricity Council Research Centre, ECRC/M801, March 1975.
48. JOHNSON, T. E., *et al.*, *MIT Solar 5 building: Initial Performance*, Dept. of Architecture, MIT, Cambridge, MA 02139, USA, October 1978.
49. BELL, G. H., DAVIDSON, J. N., and SCARBOROUGH, H., *Textbook of Physiology and Biochemistry*, 27th Edition, Livingstone, London, 1968.
50. YAGLOU, C. P., RILEY, E. C., and COGGINS, D. I., *Heating and Ventilating*, **8**, 31–5, March 1936.
51. WELLESLEY-MILLER, S., 'A retrofittable solar dehumidifier', Solar Cooling for Buildings, *Conference Proceedings*, pp. 145–50, Los Angeles, CA, USA, 6–8 February 1974.
52. DAVIES, M. G., 'A thermal model for a solar heated building', *Building Science* Supplement, Energy and Housing, 67–76, 1975.

53. McFarland, R. D., 'PASOLE: A General Simulation Program for Passive Solar Energy', Los Alamos Scientific Laboratory, LA–7433–MS, October 1978.

54. Kohler, J. T., and Sullivan, P. W., 'TEANET: A Numerical Thermal Network Algorithm for Simulating the Performance of Passive Systems on a TI–59 Programmable Calculator', Total Environmental Action Inc., Harrisville, NH03450, USA.

55. North East Solar Energy Center, USA, 'Validation Experience: Available Passive Design Programs', *Technical Report 2*, 70 Memorial Drive, Cambridge, MA 02142, USA, January 1979.

56. Glennie, W. L., 'Hand-held calculator aids for passive design', *3rd National Passive Solar Conf. Proc.*, San Jose, USA, January 1979.

57. Milbank, N. O., and Harrington Lynn, J., 'Thermal Response and the Admittance Procedure', IHVE *Symposium Proceedings*, London, 7 June 1973.

58. Olgyay, V., and Olgyay, A., *Solar Control and Shading Devices*, Princeton University Press, NJ, USA, 1975.

59. Olgyay, V., *Design with Climate*, Princeton University Press, NJ, USA, 1963.

60. Fanger, P. O., *Thermal Comfort*, McGraw–Hill Book Company, New York, USA, 1970.

61. ASHRAE, *Thermal Environmental Conditions for Human Occupancy*, Standard, 55–74, 1974.

62. Marshal, H. E., and Rugg, R. T., 'Energy conservation through life-cycle costing', *J. Architectural Education*, February 1977.

63. Chapman, P. F., 'The economic evaluation of solar energy schemes, UK section of I.S.E.S. *Conference (C12)*, The Royal Institute, London, July 1977.

64. Hopkinson, R. G., Petherbridge, P., and Longmore, J., *Daylighting*, Heinemann, London, UK, 1966.

65. Illuminating Engineering Society, UK, *Daytime Lighting in Buildings*, Illuminating Engineering Society Technical Report No. 4, July 1972.

66. Supplement to the Illuminating Engineering Society Technical Report No. 4, September 1977. Both available from the CIBS, 49 Cadogan Square, London SW1.

67. Mitchell, H. J., CAP Microsoft Ltd, *Data Logging System Review*, Vols. 1 and 2. Report prepared for Dr D. M. Bartholomew, Energy Technology Support Unit, Harwell, July 1978.

68. Private communication with Lee Dalton, LASL, USA.

69. Private communication with Blaire Hamilton, Memphremegog Group, Vermont, USA.

70. Hottel, H. C., and Woertz, B. B., 'The performance of flat plate solar heat collectors, *A.S.M.E. Transactions*, Feb. 1942.

Chapter 4

Worked Example—Presentation and Performance Analyses

4.1. DESIGN PRESENTATION

The dwelling type chosen for this worked example is similar to that shown in Fig. 4.1. This illustration differs from the more usual basic English nineteenth century row dwelling type in several minor ways: the yard should be brick paved, the wall between the entrance hall and adjacent living room (Fig. 4.1) should be of timber and plaster construction, there should be a partition and doors between the staircase and the room adjacent to it, and there is no bay window in the dining area. Besides these differences, the base dwelling for the purpose of this book is the unimproved version of that illustrated in Fig. 4.1 (indicated by dotted lines where changes were made). The walls are mostly built of brick, the slated roof has a 30° pitch and there is a basement below two-thirds of the dwelling. The basement is potentially a useful additional space but it would be expensive to waterproof and is best left to a 'do-it-yourself' extension to the dwelling. The external walls have no damp-proof course other than one or two courses of engineering brick, and the windows are single glazed and sash.

The typical urban structure in which these houses are found (Fig. 4.2) is very similar to modern planning trends in England and, being built in the nineteenth century, they are located close to city centres or shopping areas.

The room nearest the street is traditionally a self-contained sitting room with access from the entrance hall only. This room was often used only on occasions when visitors were received. The garden or yard end of the ground floor was used for unseemly domestic activities such as the outdoor toilet, the coal house and trash cans. The garden or yard was usually used for growing vegetables and possibly for keeping animals.

It is thought that a modern lifestyle is more suited to the rehabilitation design shown in Fig. 4.3, where the more leisurely living activities are given the privacy and quietness of the garden side. The garden is laid to lawn with flowered borders, and plants are introduced to the brick paved area adjacent to the kitchen. Daylighting, in the rooms on the south side of the dwelling, is improved by extending the windows down to the ground,

HOUSE WITH EXISTING TWO STOREY 'LONG' EXTENSION

Total area: 1160 sq ft.

Frontage: 17 ft.

Depth: 24 ft + 24 ft extension.

No. of persons: 5

Entrance: Dual entrance—service entrance through yard. Front door off street.

Garden: 55 ft.

Services: Gas fired boiler for hot water and central heating (radiators). (More rads can be added.) Electrical Immersion heater for hot water in summer.

Fireplaces retained.

Comments: Street environment poor — narrow pavements and streets inhibit tree planting. Well designed window boxes could compensate, and also prevent overlooking into front rooms. Daylighting at rear is very poor—hence need to enlarge windows. Existing cellar should not be wasted.

Special attention has been given to retain high quality of existing internal details (fireplace surround, door and architrave mouldings, door furniture, etc). Part of the existing extension is demolished to improve lighting conditions at rear. Rear window in living room enlarged—new rear kitchen window.

The back lane is wide and pleasant and the garden quite long. Dustbins and garages can be accommodated leaving adequate small gardens.

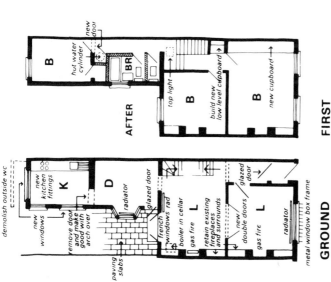

Fig. 4.1. Typical English row house plan [1].

New work is shown hatched, demolitions dotted.

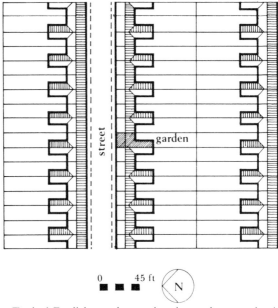

Fig. 4.2. Typical English row house site plan and assumed orientation.

avoiding the expense wherever possible of increasing the width of the openings. The living room, dining room and kitchen can all be opened on to the garden in summer and so provide comfortable living conditions. A utility room is almost a necessity; it keeps activities such as clothes washing, drying and ironing out of the living areas. The trash can is located so that there is possible access to it from both outside and inside the dwelling. The bathroom and toilet do not require windows; they can be inexpensively mechanically vented. Wash basins and cupboards are provided wherever space permits. The playroom/study is a possible spare bedroom if needed. An attempt was made to avoid openings in the wall between the two main rooms on each floor. This is a load bearing wall and consequently would be expensive to remove. As suggested in Fig. 4.1., garaging can conceivably be placed at the end of the garden.

To improve this design for solar collection and thermal performance, the changes shown in Figs 4.4, 4.5 and 4.6 are suggested. These changes were developed by means of rigorous performance testing. Alternative insulation levels, and collection window design and placement, were tested and the solution shown here evolved gradually as a result of these tests. The building has become isolated from the outdoor climate by an external

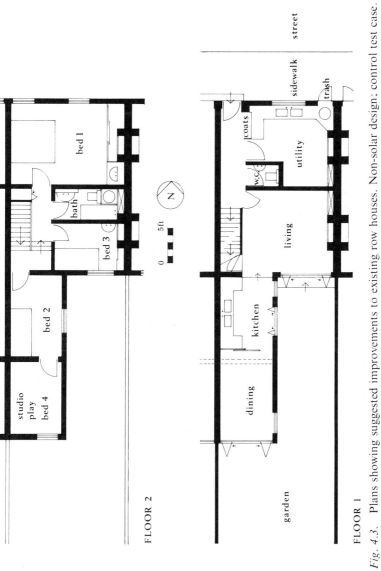

Fig. 4.3. Plans showing suggested improvements to existing row houses. Non-solar design; control test case.

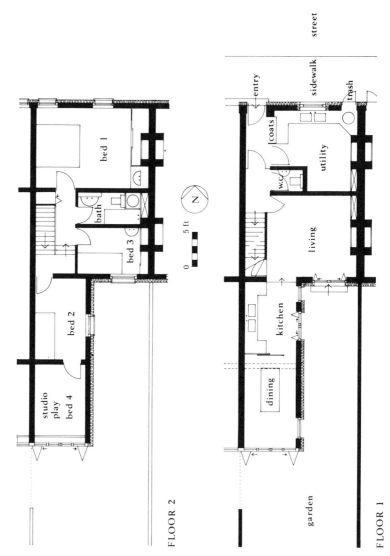

Fig. 4.4. Solar design. Plans showing suggested improvements to existing row houses.

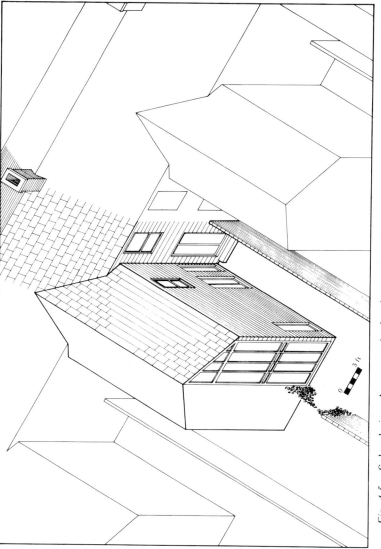

Fig. 4.5. Solar design. Axonometric showing suggested improvements to existing row houses.

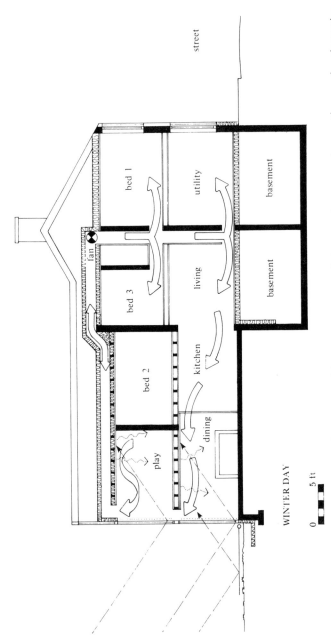

Fig. 4.6. Solar design. Section showing suggested improvements to existing row houses, and operation mode schematics.

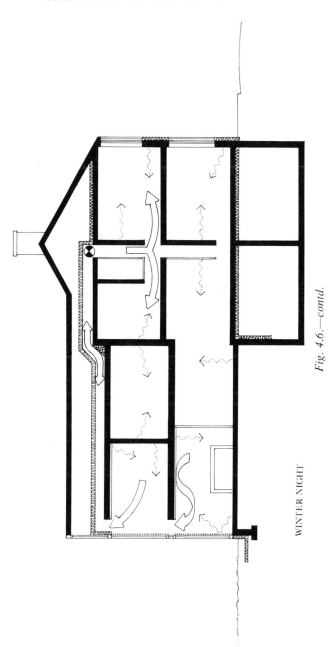

WINTER NIGHT

Fig. 4.6.—contd.

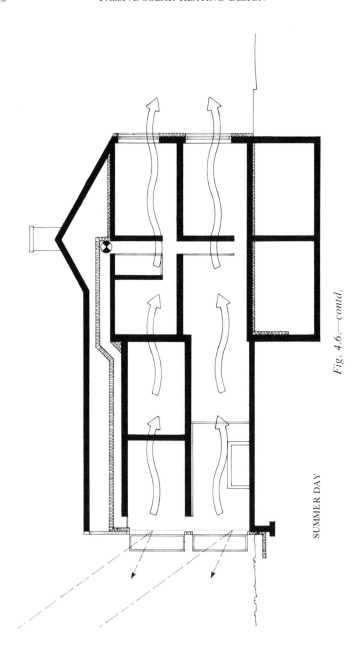

SUMMER DAY

Fig. 4.6.—contd.

nsulating layer. This layer allows the masonry external walls to be used as hermal storage mass. Solar energy collection is carried out in the dining ind playrooms. Larger collection areas were attempted but either overshadowing or the expense of changing the building's plan form ruled against them. Primary thermal mass or target area is provided by the phase change material tiles* (Sol–Ar–Tiles®) on the ceiling of the dining room and playroom. Solar modulators reflect the solar energy onto this target area. The area of the secondary thermal mass layer in the collection rooms s approximately twice that of the target and 150 mm (6 in) thick. Because he collection areas are out on a limb, a distribution system is required to move the heat to the other parts of the house. Circulating the room air in his way will have the same effect as the use of a storage bin in an active collection system. Here the masonry walls throughout the house will store any excess energy by means of convective gains from the circulating air. To avoid confusion between the function of these walls and that of the secondary thermal mass, which is in thermal radiative contact with the target area, these masonry walls shall be termed the convective thermal mass. The movement of air is triggered off by a temperature difference of 2·8 °C (5 °F) between the collection room air temperature and the living room air temperature. The delivery grilles may be hand-closed to provide control of areas to which the heat is to be delivered. On a winter evening heating is by radiation primarily, although additional heat may be stripped off the Sol–Ar–Tiles® and delivered to the spaces needing it by use of the fan. In summer the design relies on natural ventilation which will cool the building at night and leave it less susceptible than a non-solar dwelling to overheating in the day. The collection windows may be opened and the solar modulators used (or partly lowered) to reflect the solar energy away from the building. The dining room and play area then become indoor/outdoor spaces, providing a life-style unequalled by present terraced house type dwellings in England.

The following is a list of several miscellaneous design decisions taken during the course of the development of this design solution. To attempt to solve the conflict between privacy and solar energy collection, the garden wall is replaced by a thin deciduous hedge. This provides privacy when most needed, in the summer, and transmission of solar energy in the winter when privacy is less of a problem (Fig. 4.5). Return air flow is by way of gaps under the doors. The sliding screen between the kitchen and dining room is made of open mesh caning to permit air flow. Bedroom 2 is heated by way of

---

* See Chapter 3, Section 3.7 ('Thermal Mass').

thermal radiation from secondary thermal mass. The kitchen and dinin room floors are of brick to give visual continuity between indoors an outdoors and to provide additional thermal mass. All other floors ar carpeted except possibly the bath, toilet and utility rooms. The voi between the dining room and playroom is to allow air circulation. / handrail is placed in the playroom to provide protection when the window are open.

The dwelling design shown in Fig. 4.3 is used as a control case for th costing analysis discussed later in this chapter.

4.2.   PERFORMANCE ANALYSES

(a) Overheating

The data bases collected for this analysis are discussed and listed i Appendix E.

The design program presented (PDP) (Section 3.14 and Appendix C) wa extended to enable it to model the suggested solar improvements of the rov house. The changes made gave the program the ability to model more tha one type and size of collection area. This would allow modelling of, say, collection area in bedroom 3 (Fig. 4.4). Such a location for solar collectio was found to be uneconomical because the recess caused overshadowin and thus reduced solar gains. Nevertheless, this flexibility in the progran allows modelling of the dining room, with its east-facing window separately. A second alteration to the program was made. This enabled th program to model the effects of coupling the air in the collection rooms t what is, in effect, a storage bin. When the room air temperature exceed $2.8\,°C$ ($5\,°F$) above or below the temperature of the convective therma mass, the fan (Fig. 4.6) is switched on, circulating air from the collectio room to other parts of the house. When this happens, convective gains to or losses from, the convective thermal mass occur and the air temperature i decreased or increased respectively. The modelling of this is rather crude; i takes effect after the hourly collection space air temperature has bee determined and only divides the hour into eight parts. Nevertheless thes crudities would have little significant bearing on the conclusions draw from the results of the program run (Table 4.1): the room air temperatur does not overheat.

Each time the fan is switched on represents $7.5\,min$ of operating time thus the fan is on for $15\,min$ for each hour of this clear day. This is no excessive fan use for such low air temperature results and for a clear da

Table 4.1. *Results of the passive design simulation program for the solar row house rehabilitation (Fig. 4.4) in London on a clear October day*

Time of day	Start 08.00	09.00	10.00	11.00	12.00	13.00	14.00	15.00	16.00
Air temperature (°C)	18·3	19·1	19·9	20·3	21·3	21·1	20·1	19·9	21·7
Convective thermal mass temperatures (°C)	18·3	18·4	18·5	18·6	18·7	18·9	19·0	19·0	19·0
Fan switched on	—	2	2	2	2	2	2	2	0

with high outdoor air temperatures. The high room air temperature at hour 16.00 in Table 4.1 is explained by the air temperature being less than 2·8 °C above the temperature of the convective thermal mass and thus the fan not having been switched on.

Theoretically, corrections to these results would be made to allow for an increased secondary thermal mass area and also to allow for the use of Sol-Ar-Tiles® on the target area. The actual area of secondary thermal mass is approximately twice the area of the noon target (Appendix E, (a), data base 21), whereas the program assumes an area of secondary thermal mass of 0·5 times the target area. As shown in Fig. C.2 (Appendix C), this would mean an effective reduction in maximum room air temperature of about 4 °C. As discussed in Chapter 3, Section 3.7, the use of Sol-Ar-Tiles® would also significantly reduce the daily air temperature fluctuation. However, the effect of the convective thermal mass is so overpowering in determining the results shown in Table 4.1 that these additional factors would probably not significantly change the outcome of the room air temperature results.

b) Costing

The data bases used for this analysis, and the procedures for obtaining them, are listed in Appendix E.

The program used for this analysis is the Cost/Load Program (Appendix D) and the design method employed within the program is discussed in Chapter 3, Section 3.18.

To conduct a costing analysis it is first necessary to find the difference in yearly heating load between the solar and non-solar rehabilitation designs. Such heating load calculations are performed within the Cost/Load Program. Consequently all design decisions affecting the heat loss of the two proposals must be made. Such design decisions will dictate the choice of data bases; the following list deals with several of these decisions.

(i) The orientation and tilt chosen for the collecting surface is due sout and vertical. These decisions were made on the basis of several influencin factors which are discussed in Chapter 3, Section 3.3.

(ii) It is assumed that the large windows in the dining and living rooms (the non-solar design (Fig. 4.3) have movable shading devices to prever overheating in these spaces and to prevent fading of furnishings. Thes devices are assumed to be in operation for all months other than Decembe. January and February.

(iii) The colour tones used in decorating the non-collecting rooms of th solar design (Fig. 4.4) are assumed to be slightly darker than those used o the non-solar design. In England, most dwelling interiors are painted ligh colours to help compensate for the low levels of daylight during the winter This could be achieved by having a light-coloured floor, which accepts mo: of the initial daylight entering the room, and slightly darker walls an ceiling. The effect of such slightly darker tones is to increase absorption (solar radiation and thus reduce back-reflected losses.

(iv) The solar design is assumed to have 100 mm of external styrofoar insulation on all walls. This is clad with expanded metal lathing an rendered. Both the roof and suspended floor are insulated with polythene wrapped fibreglass insulation; 150 mm and 200 mm respectively. Finall the windows are all fitted with double glazing with a layer of double-side heat mirror between.

(v) The non-solar design is assumed to have 50 mm of fibreglass insulatio in the roof, single glazing to all windows, a 'Sandtex' or similar waterproc finish to the north wall, and an uninsulated ground floor. Thes specifications are in line with rehabilitation work being executed in Englan today.

(vi) The microclimate of a densely populated city will have an influenc on the heat losses of a building in such a location. Corrections for thi influence are made in the references from which the U-values of the weathe skin components are assessed (Appendix E, Table E.28). The results fror the heating load section of the Cost/Load Program runs are shown in Tabl 4.2. The price of fuel was taken to be £0·0114 per kWh, which is th approximate cost of useful energy from North Sea Gas in Englan (Appendix E, (b), data base 7). The heating load figures for the non-sola design in Table 4.2 are in good agreement with published figures b Burberry [2].

Table 4.2. *Heating load results for both solar and non-solar designs from the Cost Benefit Program runs*

Design type	Balance point temp. (°C)	Degree days (Table 3.10)	Heating load (kWh/y)	Cost of heating per year (£/y)
on-solar	15·26	2 091·38	22 089·11	251·82
olar	6·56	430·64	1 247·37	14·34

Table 4.3 shows the results of several runs of the Life-Cycle Costing ection of the Cost/Load Program. Account is taken for the economy's flation rate and the spiral rate of the increase in fossil fuel prices by using qn (26) (Section (3.18) in this section of the program. A breakdown of the otal cost of the improvements is taken mainly from *Spon's Architects' and uilders' Price Book* [3] (updated to 1979 prices from the March 1979 uilding Specification Cost Index) and is listed in Appendix E. The nprovements are assumed to last for the life of the building and thus the orrowing period is that of the normal length of a mortgage. Using a reasury Discount Rate of 5 % (Section 3.18), which would be applicable to ocal authority work, the payback period is $9\frac{1}{2}$ years. For every year after iis $9\frac{1}{2}$ years, the inhabitant will be saving over £230 per year on his heating ills. Further discussion on the values used for the fuel cost spiral rate and ie discount rate may be found in Chapter 3, Section 3.18.

The 3 % discount rate is one taken to represent a private scheme. It is seen iat this will pay for itself in $7\frac{1}{2}$ years.

Table 4.3. *Results from the Cost Benefit Program showing the time taken to recoup the additional capital expenditure involved in making the solar improvements suggested (the difference between Figs 4.3 and 4.4)*

Cost of Improvement (£)	Borrowing period (years)	Fuel cost spiral rate (%)	Discount rate (%)	Payback period (years)
2 015·00	25	4	5	9·5
			3	7·5

4.3. CONCLUSIONS

ne of the main objectives of this book has been met by showing that these nglish row houses can be successfully rehabilitated incorporating solar heating. It is to be emphasized that only a year ago such successful results were beyond any expectations. They have been made possible by the recent advances in technology by researchers at MIT and in research laboratories in the USA. The heat mirror, the Sol–Ar–Tile® and the solar modulator may individually seem to be small advances but they and other new materials in this field will revolutionize architectural thinking in the temperate climate.

The results obtained from the performance analyses, and the design proposal, seem to have implications beyond the immediate scope of this work. If, as the results of this research suggest, a ten-year payback period is attainable with improvement work in England, there is good reason to believe that the payback period on new housing there may be less. There is even more reason to believe that such a passive design approach will be applicable to the majority of underheated climates of this world, for it is well known that England suffers from little sunshine in winter.

New housing lifts many of the constraints imposed by improvement work. Orientation may be chosen, the proportion of solar collection area to floor area could be increased, the materials used for improving solar collection would be replacements of, rather than additions to, conventional materials and thus real costs would be reduced, and dwellings could be designed where natural convection is used to distribute the heat.

England has experienced a change in the structure of its society in the past 40 years. There is an increasing tendency towards a reduction in working hours per week. This has a marked effect on the life-style of the working family, placing far more emphasis on the leisure hours and the way they are spent in the home. Domestic architecture in England has, as yet, not fully responded to this change in society.

The results of this research suggest that passive solar design may be able to provide an answer to these new architectural demands. It is possible to rehabilitate row houses or build new houses in England incorporating a nurturing of life-style otherwise unthinkable because of the potentially high heating costs of a conventional design with equal amenities.

REFERENCES

1. DEPARTMENT OF THE ENVIRONMENT, UK, Area Improvement Note 4—*House Improvement and Conversion*, 1972.
2. BURBERRY, P., 'Energy: 3 design decisions', *The Architect's Journal*, Fig. 4, 343, February 1976.
3. DAVIS, BELFIELD and EVEREST, Chartered Quantity Surveyors, *Spon's Architects' and Builders' Price Book*, 102nd Edition, 1977.

Example Calculations of Heating Load Using the Conventional Degree Day and the Proposed Adjusted Degree Day Methods

These examples are conducted in imperial units because the data available for degree days at various balance point temperatures is also in imperial units (Table A.1).

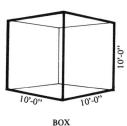

BOX

Assume that a windowless box is placed in Boston for a winter with constant internal heat gains from a television (110 W), one person (400 Btu/h), and one light bulb (60 W). The box's dimensions are 10 ft^3, the internal thermostat set point temperature is 70 °F, the heat capacity of the air is 0·018 Btu/ft^3 °F, and the infiltration rate is assumed to be $1\frac{1}{2}$ air changes per hour. Finally, walls and roof are constructed of $\frac{1}{2}$ in. gypsum, 4 × 2-in. timber stud and 1 in. external timber boarding. This has a U–value of 0·33 334 Btu/h ft^2 °F.

a. *By the conventional method:*

$$\text{Heat loss/h °F} = (U\text{–value} \times \text{Area*}) + (\text{Heat capacity} \times \text{Vol.} \times A_{\text{ch}})$$

$$= (0·33\,334 \times 500) + (0·018 \times 1000 \times 1\tfrac{1}{2})$$

$$= 193·67 \text{ Btu/h °F}$$

Therefore

$$\text{Heat loss/degree day} = 193·67 \times 24$$

$$= 4648 \text{ Btu/degree day}$$

Assumed no losses through the floor slab.

147

Table A.1.   Heating degree days as a function of balance point temperatures for Boston, Massachusetts

Balance point temp. (°F)	Heating degree days	Difference	Balance point temp. (°F)	Heating degree days	Difference
11	0	0	42	1 137	126
12	0	0	43	1 263	131
13	0	0	44	1 394	137
14	0	0	45	1 531	142
15	0	0	46	1 673	148
16	0	0	47	1 821	153
17	0	0	48	1 974	158
18	0	0	49	2 132	164
19	0	0	50	2 296	170
20	0	0	51	2 466	177
21	0	0	52	2 643	184
22	0	0	53	2 827	190
23	0	0	54	3 017	196
24	0	0	55	3 213	203
25	0	0	56	3 416	209
26	0	0	57	3 625	215
27	0	0	58	3 840	222
28	0	0	59	4 062	228
29	0	34	60	4 290	234
30	34	54	61	4 524	240
31	88	64	62	4 764	248
32	152	73	63	5 012	254
33	225	79	64	5 266	261
34	304	85	65	5 527	267
35	389	91	66	5 794	274
36	480	97	67	6 068	282
37	577	102	68	6 350	289
38	679	107	69	6 639	298
39	786	112	70	6 937	308
40	898	117	71	7 245	318
41	1 015	122	72	7 563	330

Data processed from the US National Climatic Center's Weather Tapes, Number TD 9641, *Daily Average Normal Temperatures for 379 Weather Stations*. A complete set of tables for all 379 US weather stations is obtainable from the author.

b. *By the adjusted degree day method:*

$$\text{Adjusted degree day base temp.} = 70 - \frac{\text{Incidental internal gains}}{\text{Heat loss}}$$

(from eqn (19) Section 3.11)

$$= 70 - \frac{23\,240}{4648}$$

$$= 65\,°F$$

Thus both the adjusted and non-adjusted degree day methods will give the same result of 5527 degree days. (from Table A.1).
Therefore

$$\text{Heat loss/year} = 5527 \times 4648$$

$$= 25·69\,\text{MBtu}$$

$$\text{Approximate cost at } \$14/\text{MBtu} = \$360/\text{year}$$

Now add 2 in. of expanded polystyrene insulation

$$\text{New } U\text{–value} = 0·043$$
$$\text{New heat loss/degree day} = 1169·74\,\text{Btu/day}\,°F$$

2a. *By the conventional method:*

$$\text{Heat loss} = 1169·74 \times 5527$$

$$= 6·47\,\text{MBtu/year}$$

$$\text{Approximate cost} = \$90·51/\text{year}$$

Therefore

$$\text{Calculated savings} = \$269·49$$

2b. *By the adjusted degree day method proposed here:*
(Adjusted degree day base temp.)

$$\text{Balance point temp.} = 70 - \frac{\text{Internal gains}}{\text{Heat loss}}$$

$$= 70 - \frac{23\,240}{1169·74}$$

$$= 70 - 19·9$$

$$\simeq 50\,°F$$

Number of degree days (from Table A.1) = 2296

Therefore

$$\text{Heat loss/year} = 1169\cdot74 \times 2296$$
$$= 2\cdot69\,\text{MBtu}$$
$$\text{Approximate cost} = \$37\cdot60/\text{year}$$

Therefore

$$\text{Calculated savings} = \$322\cdot40$$

The increase in savings by the use of the proposed adjusted degree day method is 20 %. This could be greater depending on the nature of the internal gains and external wall of the building analysed.

A User Manual for Passive Simulation Program (PSP)

B.1. PURPOSE OF THE PASSIVE SIMULATION PROGRAM

PSP is a finite difference thermal network program. The thermal network is undefined by the program and therefore it has great flexibility to model different passive systems. The programmer designs the network so that it resembles his solar system. PSP may be used, by means of simple modifications, to model either a heating or a cooling mode. The information supplied by the program is in terms of the temperatures within the building and the auxiliary energy introduced (if any) to maintain a certain minimum thermostat set temperature. This thermostat set temperature is an input option and may be overridden to allow the building to 'freewheel'.

B.2. PROGRAM DESCRIPTION

The PSP is run on the TI–59 (Texas Instruments) programmable calculator and PC–100A (or B) printer. It is essential to use the master module of the TI–59 when running PSP because the thermal network analysis is made possible by use of the matrix inversion library program in this module.

PSP consists of two programs: the PSP Data Storage Program and the main program, PSP. The Data Storage Program is run first to input and organize the data to be used for PSP. The Data Storage Program also finds the determinant of the large matrix and checks some inputs for error entry.

The program takes approximately 30 min for a 24 h simulation of a seven node network. A preset time increment of 1 h is assumed within the program. Smaller time increments may be used but the hour number will not identify the time increment so used.

PSP will generate wave forms for hourly outdoor temperatures and hourly solar gains per unit area (see 'Program Calculations'). It is possible to bypass this program routine and manually input the outdoor temperature and solar gains per unit area for each hour modelled.

To make design changes for rerunning PSP will necessitate altering the thermal network and therefore a rerun of the Data Storage Program will also be necessary. Second day modelling of the same conditions is easily accomplished without the need to rerun the Data Storage Program.

B.3. PROGRAM CALCULATIONS

The algorithm in PSP is based on standard procedures of unsteady state heat flow analysis [1]. If one considers a row of three nodes, the condition of the middle node is defined by the following relationship for a 1 h time step, where n denotes the node number and t the time in hours:

$$(T_n^{t+1} - T_n^t)\text{Cap}_n = \text{Con}_{n,n+1}(T_{n+1}^{t+1} - T_n^{t+1}) + \text{Con}_{n,n-1}(T_{n-1}^{t+1} - T_n^{t+1}) \tag{28}$$

$$\Delta T \times \text{Capacitance} = \text{Conductance} \times \Delta T'$$

where Cap_n = the capacitance of node $n = \text{Cap}_n/\Delta t$ for other than a 1 h time step. $\text{Con}_{n,n+1}$ = the conductance from node n to node $n+1$. $\text{Con}_{n,n-1}$ = the conductance from node n to node $n-1$. If we involve the outdoor air temperature as another possible path of heat flow and solar energy input the relationship becomes longer by these two variables being added to the right hand side:

$$+ \text{Con}_{n,a}(T_a^{t+1} - T_n^{t+1}) + I_s A_c \tau \alpha f_n$$

Making this relationship applicable to N number of nodes it becomes:

$$(T_n^{t+1} - T_n^t)\frac{\text{Cap}_n}{\Delta t} = \sum_{n=i}^{N} \text{Con}_{n,i}(T_i^{t+1} - T_n^{t-1})$$

$$+ \text{Con}_{n,a}(T_a^{t+1} - T_n^{t+1}) + I_s A_c \tau \alpha f_n \tag{29}$$

Equation (29) may be expressed as N equations [2] of:

$$\sum_{i=1}^{N} B_{ni} T_i^{t+1} = C_n \tag{30}$$

where

$$n = 1 \text{ to } N$$

and where

$$B_{ni} = \frac{Cap_n}{\Delta t} + \sum_{i=n}^{N} Con_{n,i} + Con_{n,a} \qquad \text{when } i = n$$

$$B_{ni} = -Con_{n,i} \qquad \text{when } i \neq n$$

$$C_n = \frac{Cap_n}{\Delta t} T_n^t + Con_{n,a} T_a + I_s A_c \tau \alpha f_n$$

In PSP, eqn (30) is solved by means of the master library's matrix inversion program. 'B_{ni}' is the information supplied in matrix **B** and C_n is that supplied in each row of matrix **C** relating to each node T_n.

The outside air temperature is generated by the following equation expressed in radians:

$$T_a = \frac{T_{sw}}{2}\left[\cos\left(\frac{\pi(14 - t)}{12}\right)\right] + T_{av} \qquad (31)$$

Equation (31) sets the maximum outside air temperature at 2.00 p.m. and the minimum at 2.00 a.m.

The solar gain generator uses the expression [2] (also in radians):

$$I_s = I_{max}\left[\sin\left(\frac{\pi(t - t_{sr})}{2(12 - t_{sr})}\right)\right] \qquad (32)$$

B.4. DEVELOPMENT AND TESTING

A program similar to 'TEANET' was written and tested by the author against published results of 'TEANET'. 'TEANET' has itself been through rigorous validation procedures [3]. PSP was then developed on this grounding.

B.5. LIMITATIONS

The main limitation of this program is the length of time needed to simulate a 24 h period (30 min). At least two or three consecutive clear days must be simulated to attain equilibrium temperatures within the structure of a passive solar system. If partial equilibrium temperatures are not attained erroneous results will be gained.

The second limitation is due to the limited program and storage capacity of the TI–59 calculator. No movement of energy through the air to convective storage can be easily modelled other than by assuming that there is a constant air flow. In reality this is never the case. In a thermosyphoning system the air flow rate is proportional to several factors including the incident solar energy, and in a forced air system the fan is controlled by a temperature differential, switching on and off when required.

Both limitations relate to the machine used rather than the method. If modelling of such air movement is required a larger machine should be used and the method can be taken from an understanding of the PSP listing and the methods employed in matrix inversion. A larger machine would also run this program in a fraction of the time taken on the TI–59 calculator.

Because this is a two-dimensional thermal network it is not possible to model the movement of the sun within the space and thus give a more accurate picture of the importance of placement of thermal mass.

Provided the internal routines for generating hourly outdoor temperatures and solar insolation are used there is no need to attend to the calculator and printer during a one day simulation run. The only limitations to this are that using such internal routines requires that the collection surface be either horizontal or oriented due south (so that insolation values are symmetrical about noon) and that no true profile of internal heat gains (people, appliances and other windows) can be included. To include such internal heat gains and to orient the collecting window other than due south or horizontal will involve tedious manual inputs for every hour simulated and the disguising of internal heat gains as solar gains through the collecting window by changing the contents of certain data registers every simulated hour.

Step	Description	Value entered	Press	Display
1 CHECK that Master Module is in place in calculator and mount calculator on printer				
2 CLEAR Calculator			2nd CP 2nd CMs	
3 REPARTITION Calculator		9	2nd OP 17	239·89
4 LOAD PSP DATA STORAGE PROGRAM (a) If program has not yet been keyed from listing then do this			LRN (key in as listed) LRN	
(b) Read from magnetic card			CLR (insert card)	1.
4′ RECORD PSP DATA STORAGE PROGRAM This follows step 4(a)—PSP Data Storage Program is written on to side one of a magnetic card		1	2nd Write (insert card)	1.
5 LOAD AND ORGANIZE DATA Data may be loaded from a magnetic card, if used before, or keyed in as follows. Where values are zero they should be entered to keep the data in sequence				

Step	Description	Value entered	Press	Display
		No. of nodes	A	No. of nodes
5—contd. INPUT 1 Number of nodes in network excluding the outdoor temperature node, T_A		1	B	1.
INPUT *2—(50) Matrix **B** of conductances and capacitances		B_{11}	R/S	B_{11}
		B_{21}	R/S	B_{21}
		$\mathbf{B_{31}}$	R/S	B_{31}
		B_{41}	R/S	B_{41}
		$\cdots\rightarrow$	$\cdots\rightarrow$	$\cdots\rightarrow$
		B_{71}	R/S	B_{71}
		B_{12}	R/S	B_{12}
		B_{22}	R/S	B_{22}
		$\cdots\rightarrow$	$\cdots\rightarrow$	$\cdots\rightarrow$
		B_{77}	R/S	B_{77}

B_{11}	B_{12}	B_{13}	B_{14}	B_{15}	B_{16}	B_{17}
B_{21}	B_{22}	B_{23}	B_{24}	B_{25}	B_{26}	B_{27}
B_{31}	B_{32}	B_{33}	B_{34}	B_{35}	B_{36}	B_{37}
B_{41}	B_{42}	B_{43}	B_{44}	B_{45}	B_{46}	B_{47}
B_{51}	B_{52}	B_{53}	B_{54}	B_{55}	B_{56}	B_{57}
B_{61}	B_{62}	B_{63}	B_{64}	B_{65}	B_{66}	B_{67}
B_{71}	B_{72}	B_{73}	B_{74}	B_{75}	B_{76}	B_{77}

If a mistake is made in any entry it is necessary to start from the top of that column and continue as before to the end of the matrix thus: by entering the column number (i)

Step	Description	Value entered	Press	Display
		i	B	i
		B_{2i}	R/S	B_{2i}
		B_{3i}	R/S	B_{3i}
		$\cdots\rightarrow$	$\cdots\rightarrow$	$\cdots\rightarrow$
		B_{77}	R/S	B_{77}

INPUT *51–(71)

Matrix C of node capacitance, conductance to T_A (outside temperature node) and solar impact

C_{11}	C_{12}	C_{13}
C_{21}	C_{22}	C_{23}
C_{31}	C_{32}	C_{33}
C_{41}	C_{42}	C_{43}
C_{51}	C_{52}	C_{53}
C_{61}	C_{62}	C_{63}
C_{71}	C_{72}	C_{73}

1	C	1.
C_{11}	R/S	C_{11}
C_{12}	R/S	C_{12}
C_{13}	R/S	C_{13}
2	C	2.
C_{21}	R/S	C_{21}
-->	-->	-->
C_{23}	R/S	C_{23}
3	C	3
-->	-->	-->
7	C	7
C_{71}	R/S	C_{71}
-->	-->	-->
C_{73}	R/S	C_{73}

If display flashes one of two possibilities has occurred:
either a row number other than that immediately following the row just finished has been entered
or a fourth value has been entered followed by R/S, i.e. C_{n1} has been entered without first entering n and pressing C.

* See 'Notes to Worksheet', pp. 163–165.

Step	Description	Value entered	Press	Display
5—contd.	To continue simply follow instructions above.			
	If an error is made in the figures entered then that row has to be reentered as follows:	n	2nd C'	n
	(where n = row number).	C_{n1}	R/S	C_{n1}
		C_{n2}	R/S	C_{n2}
	If after the matrix is entered it is noticed that the value of	C_{n3}	R/S	C_{n3}
	say C_{43} is incorrect then one need not reenter any values	$n+1$	C	$n+1$
	other than C_{41}, C_{42}, and C_{43} before proceeding to the next	$C_{(n+1)1}$	R/S	$C_{(n+1)1}$
	step (Check Matrix)	$\rightarrow$	$\rightarrow$	$\rightarrow$
		C_{73}	R/S	C_{73}
	CHECK MATRIX			
	This is a method of checking matrix **B** to examine whether or not the values entered were correct or incorrectly entered.	1	2nd B'	1.
	Each row is checked independently and the displayed result		R/S	Check for row 1
	for row n of matrix **B** should equal the sum of C_{n1} and C_{n2}	2	2nd B'	2.
	of matrix **C**. If this is not so there is an error in that row of		R/S	Check for row 2
	matrix **B** or **C**—but more likely the error is in matrix **B**	$\rightarrow$ 7	2nd B'	$\rightarrow$ 7.
			R/S	Check for row 7
	DETERMINANT (D) OF MATRIX B			
	This will take over 2 min to run and the displayed result will be a very large positive number		D	Determinant of matrix **B**

EXTERNAL (E) CONDITIONS

These can either be computed internally or entered as hourly values during the program run

either internal Computation:

INPUT *72 Maximum hourly solar insolation/unit area (W/m², Btu/h ft²)	I_{max}	E	I_{max}
INPUT *73 Time of sunrise (h)	t_{sr}	R/S	t_{sr}
INPUT *74 Average diurnal outdoor temperature (°C, °F)	T_d	R/S	T_d
INPUT *75 Half the diurnal outdoor temperature swing (°C, °F)	$\Delta T/2$	R/S	$\Delta T/2$
or hourly manual data input: This will tell the main program (PSP) to stop at relevant points during execution to accept both hourly outdoor temperature and hourly solar gains per unit area	CLR	2nd E'	1.

CLOCK AND THERMOSTAT

INPUT 76 Start time (usually 0) (h)	t_0	2nd D'	t_0
INPUT *77 Thermostat set point temperature (°C, °F)	T	R/S	T
INPUT 78 The capacitance of node 1 to which the thermostat is hooked (Wh/°C, Btu/°F)	Cap_1	R/S	Cap_1

Step	Description	Value entered	Press	Display
5—contd. INTERNAL START CONDITIONS INPUT *79–(85) Enter the start temperatures for each node, one at a time (°C, °F)				
			2nd A'	1.
		1	R/S	T_1
		T_1	R/S	T_2
		T_2	$\rightarrow$	$\rightarrow$
		$\xrightarrow{\ }$	R/S	T_7
		T_7		
6 RECORD DATA This will require three banks of the magnetic cards. Once this is done there will be no need to use the PSP Data Storage Program again for this set of data				
		4	INV 2nd Write (insert 1st card inverted)	4.
		3	INV 2nd Write (insert 1st card)	3.
		2	INV 2nd Write (insert 2nd card inverted)	2.
7 CLEAR program			2nd CP	
8 LOAD PSP (a) If program has not yet been keyed from listing then do this:			LRN (key in as listed) LRN	

(b) Read from magnetic card

CLR
(insert inverted card) 1.

8' This follows step 8(a)—program is written on to the inverted side of the magnetic card used to store the PSP Data Storage Program

1

2nd Write
(insert inverted card) 1.

9 SELECT PROGRAM MODE

The mode of PSP as listed is HEATING, if the COOLING mode is required the following changes must be made:

The result will be that auxiliary energy supplied will be that from an air conditioner and the thermostat set point temperature will be that of the air conditioner's thermostat

GTO 217
LRN
2nd Ins
2nd $|x|$
LRN
GTO 168
LRN
2nd Ins
INV
LRN

10 RUN PROGRAM

A seven node network will take approximately 30 min to complete a 24 h simulation. The following data are printed out for each hour modelled:

Hour number (1–24)

Outdoor air temperature for this hour (°C, °F)

A

Step	Description	Value entered	Press	Display
10—contd.				
	Solar gains per unit area for this hour (W/m², Btu/h ft²)			
	*(Auxiliary energy supplied for this hour) (Wh, Btu)			
	*(Total auxiliary energy supplied for all hours modelled including this hour) (Wh, Btu)			
	Temperature at node 1 for this hour (°C, °F)			
	Temperature at node 2 for this hour (°C, °F)			
	Temperature at node 3 for this hour (°C, °F)			
	Temperature at node 4 for this hour (°C, °F)			
	etc. for as many nodes as modelled (max. of 7).			
	To model a second or third (etc.) day of the same conditions in an attempt to achieve equilibrium, the following sequence necessary after the calculator has finished printing the results for hour 24:		24 STO	
			00	
			CLR STO	
			81	
			A	
	If hourly outdoor temperatures and solar gain values are being entered manually (see Step 5 'External Conditions') then the program will automatically stop twice during each hour modelled. The first time the program stops '1·00' will be displayed then:			1·00
	Enter this hour's outdoor temperature	T_a	R/S	
	Enter this hour's insolation per unit area	I_s	R/S	T_a

* These values are only printed on hours modelled in which there has been a demand for auxiliary energy input.

Notes to Worksheet

The units (metric or imperial) are determined by the input data only—there is no need to make changes within the program code.

INPUT 2–(50). *Matrix* **B**

(Note: all entries must be rounded to the nearest whole number.)

As seen in eqn (30) wherever B_{ni} has subscripts of equal values, i.e. B_{11}, B_{22}, B_{33}, etc. the value placed at this position in the matrix is the capacitance (specific heat × density × volume) of node n plus the sum of the conductances (area × U-value) of all paths of heat flow to and from node n. In all other cases of B_{ni} the values placed in the matrix are the *negative* values of the conductance from node n to node i. Therefore the value one places in say position B_{24} is the negative value of the conductance from node 2 to node 4. The total number of rows and of columns is always equal to the number of nodes being used in the thermal network. Thus if, for example, six nodes are being used the matrix will contract from its maximum of 7 × 7 to 6 × 6 and only the 36 variables are entered into the calculator.

Wherever parentheses occur around bold numbers this means that, because there is an option of how many nodes are used in the network, there may not be that number of entries.

Capacitance: Wh/°C, Btu/°F. Conductance: W/°C, Btu/h°F.

INPUT 51–(71). *Matrix* **C.**

(Note: all entries must be rounded to the nearest whole number.)

Matrix C is not treated as a true matrix by the program but dealing with the data as a matrix simplifies the problem of inputting it into the calculator.

The type of value placed in the position C_{ni} varies with the subscript i (from 1 to 3). The values always relate to node n in the thermal network (from 1 to 7).

C_{n1} ...this is the capacitance of node n. *Note:* this number must not exceed four digits.

C_{n2} ...this is the conductance from node n to the outside air temperature. *Note:* this number must not exceed three digits.

C_{n3} ...this is a number which, when multiplied by the hourly incident solar energy on a vertical surface, produces the total solar impact on node n. It therefore consists of

$$A \times \tau \times \alpha \times f_n$$

where A = the area of collection window (m^2, ft^2),

τ = the average transmission of the transparent membrane used in the collection window,

α = the absorptance (due to the surface colour) of the plane represented by node n (see Section C.7, step 8),

f = the fraction of the total incoming solar energy that falls on the plane represented by node n (0 to 1).

Note: this number (C_{n3}) must not exceed three digits.

Limitations are imposed on the number of digits allowed as entries for C_{n1}, C_{n2} and C_{n3} because these numbers have to be stacked into one data storage register in the calculator. If the values do exceed these specified number lengths then both matrices **C** and **B** will have to be divided by a number which brings matrix **C** entries into line with the specified lengths.

INPUT 72. This is the maximum hourly clear day solar insolation on a south-facing surface of tilt determined by the designer. Tables [4, 5] will provide this information related to the month chosen for modelling. (W/m^2, Btu/h ft^2.)

INPUT 73. The time of sunrise may be calculated from the following formula [2]:

$$t_{sr} = 12 - \left(\frac{\pi I}{4 \times I_{max}} \right) \tag{33}$$

where I = the total insolation over the clear day (Wh/m^2, Btu/ft^2), I_{max} = the maximum hourly solar insolation (W/m^2, Btu/h ft^2) and where t_{sr} is rounded to the nearest half hour.

INPUT 74. The average diurnal outdoor temperature for the time of year chosen to model. This is calculated by dividing the sum of the maximum and the minimum daily normal outdoor temperature (both for the month modelled) by two.

INPUT 75. The temperature swing is found by subtracting the minimum normal daily outdoor temperature from the maximum normal daily outdoor temperature (both for the month modelled). This is then halved to provide the data entry required.

INPUT 77. If there is no thermostat, i.e. the building is to be modelled while allowing it to 'freewheel', a number should be entered here which will allow this to happen. For example, if the program is being used in its heating node then the thermostat setting would be placed at 5°C (41°F) under the assumption that the indoor temperature will not fall below this point. In the cooling mode 45°C (113°F) may be used as the thermostat set point temperature. Otherwise the usual thermostat set point temperatures for heating and cooling are between 18·3°C and 22°C (65°F and 72°F).

NPUT 79–(85). The number of start temperatures entered should equal the number of nodes used in the thermal network. These temperatures should be around 15·5°C (60°F) for the first day modelled. The node (node T_1) representing the room air temperature attached to a 'useful' thermostat will have a start temperature equal to the set point temperature of the thermostat.

LISTING OF : DATA STORAGE PROGRAM FOR PSP	
PROGRAMMER : R.M. LEBENS	PAGE : 1
For use on a TI-59 calculator plus printer	

LOC	CODE	KEY	LOC	CODE	KEY	LOC	CODE	KEY
000	76	LBL	052	80	80	104	43	RCL
001	11	A	053	55	÷	105	07	07
002	36	PGM	054	01	1	106	42	STO
003	02	02	055	00	0	107	03	03
004	11	A	056	45	YX	108	25	CLR
005	91	R/S	057	03	3	109	42	STO
006	76	LBL	058	95	=	110	05	05
007	12	B	059	74	SM*	111	76	LBL
008	36	PGM	060	03	03	112	60	DEG
009	02	02	061	43	RCL	113	73	RC*
010	12	B	062	80	80	114	01	01
011	76	LBL	063	99	PRT	115	44	SUM
012	67	EQ	064	91	R/S	116	05	05
013	91	R/S	065	42	STO	117	43	RCL
014	36	PGM	066	80	80	118	07	07
015	02	02	067	55	÷	119	44	SUM
016	91	R/S	068	01	1	120	01	01
017	61	GTO	069	00	0	121	97	DSZ
018	67	EQ	070	45	YX	122	03	03
019	76	LBL	071	06	6	123	60	DEG
020	13	C	072	95	=	124	43	RCL
021	42	STO	073	74	SM*	125	05	05
022	03	03	074	03	03	126	91	R/S
023	43	RCL	075	43	RCL	127	76	LBL
024	81	81	076	80	80	128	14	D
025	85	+	077	99	PRT	129	36	PGM
026	01	1	078	91	R/S	130	02	02
027	95	=	079	76	LBL	131	13	C
028	32	X!T	080	70	RAD	132	91	R/S
029	43	RCL	081	94	+/-	133	76	LBL
030	03	03	082	34	ΓX	134	15	E
031	22	INV	083	91	R/S	135	42	STO
032	67	EQ	084	76	LBL	136	87	87
033	70	RAD	085	18	C'	137	43	RCL
034	76	LBL	086	42	STO	138	87	87
035	80	GRD	087	03	03	139	99	PRT
036	43	RCL	088	42	STO	140	91	R/S
037	03	03	089	81	81	141	42	STO
038	42	STO	090	61	GTO	142	88	88
039	81	81	091	80	GRD	143	99	PRT
040	07	7	092	76	LBL	144	91	R/S
041	00	0	093	17	B'	145	42	STO
042	44	SUM	094	42	STO	146	89	89
043	03	03	095	03	03	147	99	PRT
044	43	RCL	096	85	+	148	91	R/S
045	81	81	097	07	7	149	42	STO
046	91	R/S	098	95	=	150	86	86
047	72	ST*	099	42	STO	151	99	PRT
048	03	03	100	01	01	152	91	R/S
049	99	PRT	101	43	RCL	153	76	LBL
050	91	R/S	102	03	03	154	16	A'
051	42	STO	103	91	R/S	155	42	STO

LISTING OF : DATA STORAGE PROGRAM FOR PSP

| PROGRAMMER : R.M.LEBENS | PAGE : 2 |

For use on a TI – 59 calculator plus printer

LOC	CODE	KEY
156	05	05
157	76	LBL
158	38	SIN
159	91	R/S
160	42	STO
161	78	78
162	43	RCL
163	07	07
164	33	X²
165	95	=
166	50	I×I
167	85	+
168	07	7
169	95	=
170	42	STO
171	01	01
172	43	RCL
173	05	05
174	32	X:T
175	43	RCL
176	07	07
177	22	INV
178	77	GE
179	38	SIN
180	76	LBL
181	89	∏
182	01	1
183	44	SUM
184	01	01
185	73	RC*
186	01	01
187	22	INV
188	77	GE
189	89	∏
190	43	RCL
191	07	07
192	44	SUM
193	01	01
194	01	1
195	44	SUM
196	05	05
197	43	RCL
198	78	78
199	72	ST*
200	01	01
201	99	PRT
202	61	GTO
203	38	SIN
204	91	R/S
205	76	LBL
206	19	D'
207	42	STO

LOC	CODE	KEY
208	81	81
209	75	-
210	02	2
211	04	4
212	50	I×I
213	42	STO
214	00	00
215	43	RCL
216	81	81
217	99	PRT
218	91	R/S
219	42	STO
220	84	84
221	99	PRT
222	91	R/S
223	42	STO
224	83	83
225	99	PRT
226	91	R/S
227	76	LBL
228	10	E'
229	01	1
230	42	STO
231	87	87
232	91	R/S

LISTING OF : PASSIVE SIMULATION PROGRAM (PSP)	
PROGRAMMER : R.M.LEBENS	PAGE : 1
For use on a TI – 59 calculator plus printer	

LOC	CODE	KEY	LOC	CODE	KEY	LOC	CODE	KEY
000	76	LBL	051	54	)	102	45	Y×
001	11	A	052	65	×	103	03	3
002	70	RAD	053	89	π	104	95	=
003	58	FIX	054	55	÷	105	42	STO
004	02	02	055	02	2	106	78	78
005	01	1	056	95	=	107	59	INT
006	44	SUM	057	38	SIN	108	65	×
007	81	81	058	65	×	109	43	RCL
008	01	1	059	43	RCL	110	85	85
009	32	X:T	060	87	87	111	95	=
010	43	RCL	061	95	=	112	74	SM*
011	87	87	062	42	STO	113	01	01
012	67	EQ	063	80	80	114	43	RCL
013	50	I×I	064	22	INV	115	78	78
014	01	1	065	77	GE	116	22	INV
015	04	4	066	28	LOG	117	59	INT
016	75	-	067	76	LBL	118	65	×
017	43	RCL	068	39	COS	119	01	1
018	81	81	069	01	1	120	00	0
019	95	=	070	36	PGM	121	45	Y×
020	65	×	071	02	02	122	03	3
021	89	π	072	16	A'	123	65	×
022	55	÷	073	43	RCL	124	43	RCL
023	01	1	074	07	07	125	80	80
024	02	2	075	42	STO	126	95	=
025	95	=	076	03	03	127	74	SM*
026	39	COS	077	07	7	128	01	01
027	65	×	078	00	0	129	01	1
028	43	RCL	079	42	STO	130	44	SUM
029	86	86	080	79	79	131	01	01
030	85	+	081	76	LBL	132	97	DSZ
031	43	RCL	082	43	RCL	133	03	03
032	89	89	083	01	1	134	43	RCL
033	95	=	084	44	SUM	135	76	LBL
034	42	STO	085	79	79	136	52	EE
035	85	85	086	71	SBR	137	25	CLR
036	25	CLR	087	30	TAN	138	36	PGM
037	32	X:T	088	65	×	139	02	02
038	43	RCL	089	73	RC*	140	15	E
039	81	81	090	79	79	141	43	RCL
040	75	-	091	59	INT	142	81	81
041	43	RCL	092	95	=	143	99	PRT
042	88	88	093	72	ST*	144	43	RCL
043	95	=	094	01	01	145	85	85
044	55	÷	095	73	RC*	146	99	PRT
045	53	(	096	79	79	147	43	RCL
046	01	1	097	22	INV	148	80	80
047	02	2	098	59	INT	149	99	PRT
048	75	-	099	65	×	150	01	1
049	43	RCL	100	01	1	151	36	PGM
050	88	88	101	00	0	152	02	02

LISTING OF: PASSIVE SIMULATION PROGRAM (PSP)

PROGRAMMER: R.M. LEBENS | PAGE: 2

For use on a TI-59 calculator plus printer

LOC	CODE	KEY
153	16	A'
154	71	SBR
155	30	TAN
156	43	RCL
157	07	07
158	75	-
159	01	1
160	95	=
161	42	STO
162	03	03
163	43	RCL
164	78	78
165	32	X:T
166	43	RCL
167	84	84
168	77	GE
169	44	SUM
170	76	LBL
171	34	ΓX
172	73	RC*
173	01	01
174	99	PRT
175	76	LBL
176	45	YX
177	01	1
178	44	SUM
179	01	01
180	71	SBR
181	30	TAN
182	99	PRT
183	97	DSZ
184	03	03
185	45	YX
186	97	DSZ
187	00	00
188	11	A
189	91	R/S
190	76	LBL
191	50	IXI
192	91	R/S
193	42	STO
194	85	85
195	91	R/S
196	42	STO
197	80	80
198	61	GTO
199	39	COS
200	76	LBL
201	30	TAN
202	73	RC*
203	01	01

LOC	CODE	KEY
204	42	STO
205	78	78
206	92	RTN
207	76	LBL
208	44	SUM
209	43	RCL
210	84	84
211	72	ST*
212	01	01
213	75	-
214	43	RCL
215	78	78
216	95	=
217	65	×
218	43	RCL
219	83	83
220	95	=
221	44	SUM
222	82	82
223	99	PRT
224	43	RCL
225	82	82
226	99	PRT
227	61	GTO
228	34	ΓX
229	76	LBL
230	28	LOG
231	25	CLR
232	42	STO
233	80	80
234	61	GTO
235	39	COS

B.7.   PSP WORKED EXAMPLE

The example selected is a large house with a south-facing conservatory i
London. The month chosen is March. It is assumed that the conservatory i
unvented and closed off from the house by means of windows.

Step 1
Draw a thermal network to represent the heat flow within the solar system
Figure B.1 is such a thermal network. The 'Con's are conductances an

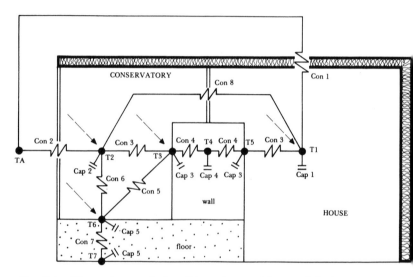

Fig. B.1.   Thermal network describing a house with a south-facing conservatory
The conservatory is unvented and closed off from the house by means of windows

'Cap's are capacitances. There is thermal mass on the floor (200 mm, 8-in
thick concrete) and in the walls (300-mm, 1-ft thick brickwork). Con 3 and
Con 6 represent the convection to or from these thermal masses whereas
Con 5 represents infrared radiation heat transfer between them. Solar
energy, represented by dashed arrows, is assumed to fall, in varying
amounts, on to nodes T_1, T_2, T_3 and T_6. There is assumed to be negligible
heat loss from the concrete conservatory floor to the outside. It will be
noticed that T_A, the outdoor temperature node, is not one of the seven
allowable nodes. T_1 must be used for the room air temperature where there

s a thermostat setting—for it is only T_1 that has the capability of being acted upon by auxiliary energy supply.

Step 2

Determine the capacitances associated with each node and the conductances between the nodes. Wherever there is no direct line of heat flow between two nodes the value of the conductance becomes zero. For example the conductance between nodes T_3 and T_5 is zero.

When the conductance is to T_A then the rate of heat loss due to infiltration must be included (number of air changes per hour × volume of air in space × 0·34, where 0·34 (Wh/m^3°C) is the specific heat of the air (0·018 Btu/ft^3°F)).

The only problem in determining the capacitances and conductances is in 'Area' and 'Volume'. The 'Area' means the total area of the building element or structure through which the conductance acts. 'Volume' means the total volume of the layer represented by that node—for example the brick wall in Fig. B.1 has three nodes and therefore three capacitances. All three nodes represent the same area of wall. The thickness of wall is divided between the three nodes. Because there are two conductances, half of Con 4 is associated with node T_3 and the other half with T_4. Thus node T_4 will have a layer thickness of 150 mm associated with it and nodes T_3 and T_5 will each have an associated layer thickness of 75 mm.

The following conductances and capacitances are used in this example:

Con 1 = 545 W/°C, total fabric + infiltration heat losses from house
Con 2 = 241 W/°C, total fabric + infiltration heat losses from conservatory
Con 3 = 168 W/°C, convection from wall
Con 4 = 168 W/°C, conduction through 150 mm of brick wall
Con 5 = 102 W/°C, radiation between thermal masses
Con 6 = 57 W/°C, convection from floors
Con 7 = 160 W/°C, conduction through 200 mm of concrete floor
Con 8 = 175 W/°C, heat flow through separating windows

Cap 1 = 1000 Wh/°C, house air and surfaces
Cap 2 = 200 Wh/°C, conservatory air
Cap 3 = 1450 Wh/°C, brickwork 75 mm
Cap 4 = 2900 Wh/°C, brickwork 150 mm
Cap 5 = 920 Wh/°C, concrete 200 mm

Step 3

From the above information we can construct Matrix **B**.

B_{11}	B_{12}	B_{13}	B_{14}	B_{15}	B_{16}	B_{17}
1888	−175	0	0	−168	0	0
B_{21}	B_{22}	B_{23}	B_{24}	B_{25}	B_{26}	B_{27}
−175	841	−168	0	0	−57	0
B_{31}	B_{32}	B_{33}	B_{34}	B_{35}	B_{36}	B_{37}
0	−168	1888	−168	0	−102	0
B_{41}	B_{42}	B_{43}	B_{44}	B_{45}	B_{46}	B_{47}
0	0	−168	3236	−168	0	0
B_{51}	B_{52}	B_{53}	B_{54}	B_{55}	B_{56}	B_{57}
−168	0	0	−168	1786	0	0
B_{61}	B_{62}	B_{63}	B_{64}	B_{65}	B_{66}	B_{67}
0	−57	−102	0	0	1239	−160
B_{71}	B_{72}	B_{73}	B_{74}	B_{75}	B_{76}	B_{77}
0	0	0	0	0	−160	1080

Matrix **B**

Step 4

Calculate the solar impact on each node.

The number entered in Matrix **C**, positions C_{n3} is defined by the following equation:

$$C_{n3} = \text{area} \times \text{transmission} \times \text{absorptance} \times f_n$$

where area = area of collection window = $145\,\text{m}^2$,

transmission = 0·6 for heat mirror + double glazing,

absorptance = 0·9 for all nodes because the reflected solar energy is accounted for in f_n because solar energy reflected off one surface is absorbed by another (only 10% is being back reflected out through the collection window),

   f_n = 0·65 for walls (T_3) 0·18 for the conservatory floor (T_6), 0·0 for conservatory furnishings (T_2), and 0·1 for house furnishings through the separating windows (T_1).

Thus the values for C_{n3} become: $C_{13} = 8$

$$C_{23} = 5$$
$$C_{33} = 51$$
$$C_{43} = 0$$
$$C_{53} = 0$$
$$C_{63} = 14$$
$$C_{73} = 0$$

Step 5

Construct Matrix **C**.

C_{11} 1000	C_{12} 545	C_{13} 8
C_{21} 200	C_{22} 241	C_{23} 5
C_{31} 1450	C_{32} 0	C_{33} 51
C_{41} 2900	C_{42} 0	C_{43} 0
C_{51} 1450	C_{52} 0	C_{53} 0
C_{61} 920	C_{62} 0	C_{63} 14
C_{71} 920	C_{72} 0	C_{73} 0

Matrix **C**

C_{n1} values are taken directly from the capacitances of each node.
C_{n2} values are the conductances from nodes to the outdoor temperature node (T_A).
C_{n3} values are as calculated in Step 4.

Step 6

External conditions:

Maximum hourly insolation for March $= 700 \, \text{W/m}^2$

Total daily insolation(I_{tot}) $= 4830 \, \text{W/m}^2$

$$\text{Time of sunrise} = 12 - \left(\frac{\pi \times I_{tot}}{4 \times I_{max}} \right)$$

$$= 12 - \left(\frac{\pi \times 4830}{4 \times 700} \right)$$

$$= 6 \cdot 58 \simeq 6 \cdot 5 \ (6.30 \, \text{a.m.})$$

Average diurnal outdoor temperature $= 6 \cdot 75 \, °\text{C}$

Half the diurnal outdoor temperature swing $= 1 \cdot 5 \, °\text{C}$

Clock and thermostat:

Start time $= 0$

Thermostat set point temperature $= 18 \, °\text{C}$

Capacitance of node 1 $= 1000$

Start temperatures assumed:

$T_1 = 18\,°C$
$T_2 = 15\,°C$
$T_3 = 15\,°C$
$T_4 = 15\,°C$
$T_5 = 15\,°C$
$T_6 = 15\,°C$
$T_7 = 15\,°C$

Step 7

Enter data and run PSP—the following results will be printed:

Contents of data registers before PSP execution

Contents	Register	Contents	Register
24.	00	-10.29601395	46
70.	01	-1.808538393	47
62.	02	1228.063291	48
63.	03	-.1302864447	49
7.	04	0.	50
8.	05	0.	51
2.1307562 22	06	0.	52
7.	07	0.	53
1888.	08	0.	54
-0.092690678	09	-160.	55
0.	10	1059.154169	56
0.	11	1.	57
-.0889830508	12	.2.	58
0.	13	3.	59
0.	14	4.	60
-175.	15	5.	61
824.7791314	16	6.	62
-.2036908957	17	7.	63
0.	18	18.	64
-.0188802472	19	15.	65
-0.069109411	20	15.	66
0.	21	15.	67
0.	22	15.	68
-168.	23	15.	69
1853.77993	24	15.	70
-0.090625644	25	1000.545008	71
-.0017110346	26	200.241005	72
-.0612857973	27	1450.000051	73
0.	28	2900.	74
0.	29	1450.	75
0.	30	920.000014	76
-168.	31	920.	77
3220.774892	32	15.	78
-.0522506103	33	0.	79
-.0031967506	34	0.	80
0.	35	0.	81
-168.	36	0.	82
-15.5720339	37	1000.	83
-3.171881532	38	18.	84
-168.2874538	39	0.	85
1761.958294	40	1.5	86
-.0010264366	41	700.	87
0.	42	6.5	88
0.	43	6.75	89
-57.	44		
-113.6103811	45		

Results of the example PSP run

	Hour 1	Hour 2	Hour 3	Hour 4	Hour 5	Hour 6	Hour 7	Hour 8	Hour 9	Hour 10	Hour 11	Hour 12
Hour	1.00	2.00	3.00	4.00	5.00	6.00	7.00	8.00	9.00	10.00	11.00	12.00
T_a	5.30	5.25	5.30	5.45	5.69	6.00	6.36	6.75	7.14	7.50	7.81	8.05
Solar	0.00	0.00	0.00	0.00	0.00	0.00	99.62	290.79	458.40	588.88	671.65	700.00
Auxil.	4517.03	4623.83	4643.85	4616.12	4553.23	4462.88	3796.10	2512.95	1196.03			
Σ Auxil.	4517.03	9140.86	13784.70	18400.82	22954.04	27416.92	31213.02	33725.97	34922.00			
T_1	18.00	18.00	18.00	18.00	18.00	18.00	18.00	18.00	18.00	18.02	19.05	20.32
T_2	11.83	10.97	10.70	10.63	10.64	10.71	12.21	15.93	20.91	26.43	31.85	36.68
T_3	14.71	14.39	14.11	13.87	13.67	13.51	16.29	24.18	35.64	49.03	62.69	75.08
T_4	14.98	14.93	14.87	14.80	14.72	14.64	14.71	15.20	16.25	17.91	20.15	22.85
T_5	14.86	14.72	14.61	14.51	14.43	14.37	14.38	14.56	14.93	15.50	16.27	17.27
T_6	14.83	14.63	14.43	14.25	14.09	13.95	15.28	19.34	25.61	33.42	41.96	50.39
T_7	14.97	14.92	14.85	14.76	14.66	14.56	14.66	15.36	16.88	19.33	22.68	26.79

13.00	17.00	21.00
8.20	7.81	6.36
671.65	99.62	0.00
21.36	19.26	1896.16
40.38	37.34	40239.57
84.82	81.35	18.00
25.84	36.16	26.45
18.46	23.32	54.00
57.90	66.49	38.70
31.40	48.09	25.65
		55.49
		53.15
14.00	18.00	
8.25	7.50	
588.88	0.00	22.00
21.89	391.99	6.00
42.49	35313.98	0.00
90.85	18.00	2115.18
28.90	33.41	42354.75
19.77	72.20	18.00
63.76	37.40	24.92
36.19	24.11	50.01
	63.36	38.63
	50.35	25.95
15.00		53.34
8.20		53.18
458.40	19.00	
21.74	7.14	
42.73	0.00	23.00
92.43	1378.24	5.69
31.79	36692.22	0.00
21.08	18.00	2305.68
67.37	30.50	44660.43
40.81	64.83	18.00
	38.17	23.63
	24.73	46.70
16.00	60.48	38.40
8.05	51.85	26.16
290.79		51.37
20.85		52.91
40.99	20.00	
89.23	6.75	
34.28	0.00	24.00
22.30	1651.19	5.45
68.34	38343.41	0.00
44.89	18.00	2464.00
	28.27	47124.43
	58.87	18.00
	38.57	22.55
	25.24	43.94
	57.87	38.06
	52.74	26.28
		49.57
		52.41

REFERENCES

1. KREITH, F., *Principles of Heat Transfer*, 3rd Edition, Intext Press Inc., New York, 1973.
2. KOHLER, J. T. and SULLIVAN, P. W., 'TEANET: A Numerical Thermal Network Algorithm for Simulating the Performance of Passive Systems on a TI–59 Programmable Calculator', Total Environmental Action Inc., Harrisville, NH 03450, USA.
3. NORTH EAST SOLAR ENERGY CENTER, USA, 'Validation Experience: Available Passive Design Programs', *Technical Report 2*, 70 Memorial Drive, Cambridge, MA 02142, USA, January 1979.
4. IHVE, 'Guide A6', *Solar Data*, 1975.
5. ASHRAE, *Handbook of Fundamentals*, 1977.

Appendix C

A User Manual for Version 1 of Passive Design Program (PDP1)

C.1. PURPOSE OF THE PASSIVE DESIGN PROGRAM

PDP is a simple method for determining the maximum room air temperature to be expected in a south-oriented direct heat-gain passively solar heated space on a clear day at the equinoxes (21 March or 23 September).

C.2. PROGRAM DESCRIPTION

The PDP is run on the TI–59 (Texas Instruments) programmable calculator. The PCM–100A printer is not required for the program as it is listed, but the program may be easily adapted to facilitate the use of this printer (see 'PDP Worksheet').

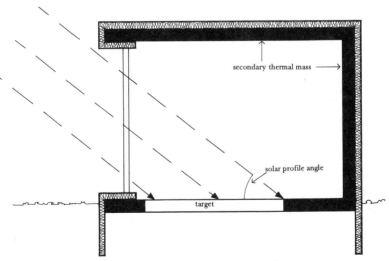

Fig. C.1. A direct gain passively heated space.

179

PDP consists of the main program plus a Data Storage Program. Th Data Storage Program is run first to input all data describing the sola system. Then PDP is run using this data. The program takes $2\frac{1}{2}$ min to ru and provides the user with eight, hourly room air temperatures and a figur for the number of hours of carry-through of the system after such a clea day of insolation. None of the data being modelled are changed durin program execution. This allows comparative modelling of the effects o design alterations without repeating data entry.

C.3.   PROGRAM DEVELOPMENT AND TESTING

PDP was developed at MIT by means of side-by side testing of a larg number of simulation runs of this program and a numerical finite differenc analysis program 'THERMAL' written by T. E. Johnson [1]. The majo variables were factored out and a simple approximation method resulte which, together with the use of graphs of the influence of the majo variables, obtained maximum daily temperature results which were withir $1\,°C\,(2\,°F)$ of the same results from Johnson's program and from actual dat on various buildings.

Johnson's Passive Simulation Program is, in structure, similar to PDP ir that it also models the equinoxes and uses the vertically directed componen of the hourly solar gains. It assumes that the target area and secondar thermal mass layer are parallel and opposite one another. Radiative and convective thermal exchange between the surfaces and room air are modelle separately with values of $5\cdot1\,W/m^2\,°C\,(0\cdot9\,Btu/h\,ft^2\,°F)$ for radiation and $2\cdot8\,W/m^2\,°C\,(0\cdot5\,Btu/h\,ft^2\,°F)$ for convection. The same values are used fo both surfaces irrespective of their position in the room. 'Thermal' models a 48 h period so that one clear and one overcast day may be simulated.

The building used as a test case was one which was long enough to make the effects of solar radiation falling on the side walls negligible, and thus the fraction of noon target illuminated was assumed to be $1\cdot0$ for all 8 h. 'Thermal' was set up so that the first 8 h being modelled were from 8.00 a.m. to 4.00 p.m. An example of the data modelled has been published [2, 3]. A target slab thickness of 6 in was chosen because, although not the optimum (9 in) thickness, it is closer to present building practice than the optimum (Fig. 3.16, Section 3.7). Both the target and secondary thermal mass were assumed to be concrete.

It was soon discovered, by altering one variable at a time, that both the area and thickness of the secondary thermal mass were the two major,

reviously unrecognised influences on the room air temperatures. An
ncrease in the thickness of the secondary thermal mass, without changing
he area, revealed a thickness beyond which there was no reduction in room
ir temperature fluctuation (see Fig. 3.16). This thickness is termed the
urning point thickness. It was also found that there is a different turning
oint thickness for each ratio of secondary thermal mass to primary
hermal mass (see Fig. 3.17 Section 3.7).

Two variable influences on the room air temperature fluctuation was one
ariable too many if a useful design method were to be produced.
Fortunately, the turning point thicknesses of secondary thermal mass areas
2–6 in—see Fig. 3.17) are easily attainable from materials in common use
n the construction industry (e.g. concrete blocks). If a graph could be
rawn showing room air temperatures against proportional areas of
econdary thermal mass (using the turning point thicknesses associated
vith each proportional area) then essentially the unknown variable effect of
he thickness of secondary thermal mass is factored out. Figures 3.18 and
.19 (Section 3.7) are both this type of graph, showing underheated and
overheated conditions respectively. There is almost no difference in the
magnitude of the effect on the room air temperature when the proportional
area of secondary thermal mass is changed in both the underheated and
overheated building designs. The only difference is that the change is seen in
he morning minimum temperatures in the underheated design, but the
change occurs to the maximum room air temperature with overheated
design (Figs. 3.18 and 3.19).

The result of this is that only one of these previously unrecognized
variables is now of enormous importance: the proportional area of
secondary thermal mass. This can be dealt with by having a design tool
which assumes a set proportional area of secondary thermal mass with its
urning point thickness. Then for any other areas of secondary thermal
mass the maximum room temperature can be corrected using the
differences on either Fig. 3.18 or Fig. 3.19. Since this is a design tool to help
guard against overheating one will usually be looking at an overheated
rather than an underheated situation and thus it is always the maximum
room air temperature which will be corrected.

Testing began on developing such a design tool by setting both (PDP and
Thermal') programs up on the same computer and running them
alternately. The area of secondary thermal mass in Johnson's program was
set at 0·5 times the area of the noon target, and the thickness of this
secondary thermal mass layer at 120 mm (4·8 in) (the turning point
thickness for that proportional area). The results were that maximum room

air temperatures, of within 1 °C (2 °F) of those from Johnson's program, were obtained with PDP by setting the thermal transfer coefficient between the target and the room air to 3·86 W/m² °C (0·68 Btu/h ft² °F), and by increasing the surface temperature of the target slab by a factor of 5·3. These are the only two fudge factors within the program and buried within them is a weighting factor which allows for thermal radiation to, and convective heat loss from, the area of secondary thermal mass (which is 0·5 times the target area). Results were tested with data from Albuquerque, Boston and London, and the same ±1 °C accuracy resulted. Results were then tested against clear day indoor temperatures obtained from the Wright house [4], MIT Solar 5 building (before installation of the Sol–Ar–Tiles®) and LASL test cell No 6 [5] and all simulations gave the same ±1 °C accuracy.

C.4.   PROGRAM CALCULATIONS

These are described extensively elsewhere [2, 3, 3a] and will not be repeated here.

C.5.   LIMITATIONS

It is necessary to stress that this program is not a true simulation tool, its only result being the maximum room air temperature of a south-oriented direct gain, passively heated space at the equinoxes. Thus it will give no auxiliary heating requirements nor follow a temperature curve equivalent to that which will actually occur within the modelled space.

Because of the stringent conditions under which this program was developed it is thought that the use of a material other than concrete for the secondary thermal mass would be expected to give different values of maximum room air temperatures. Testing of PDP has so far shown that if such materials are changed the accuracy of the results remains within 1 °C (2 °F) (the Wright house has Adobe thermal masses). A change of thickness or material of the primary thermal mass would not be expected to give different results because these factors are taken into account by the resistance split fraction (see Chapter 3, Section 3.7).

The user is discouraged from modelling any time of year other than the equinoxes because a deeper understanding of the program calculations is needed to do so and fudge factors become necessary.

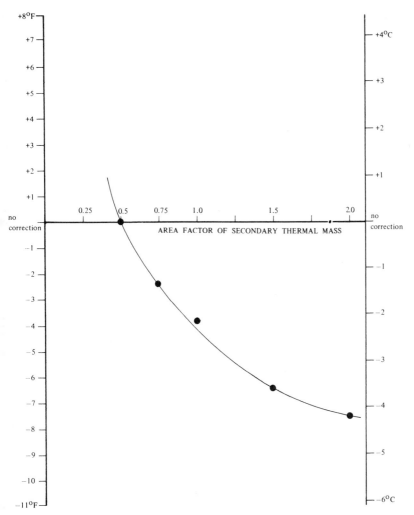

Fig. C.2. Degree Fahrenheit and Celsius corrections to be applied to maximum room air temperature results from PDP. Noon target area × area factor of secondary thermal mass = area of secondary thermal mass.

C.6. SECONDARY THERMAL MASS AND THE NEED FOR PROGRAM RESULT CORRECTIONS

True secondary thermal mass is defined as a massive element which is in longwave radiative contact with the target area, but which is never in direct sunlight (see Fig. C.1). This definition does have exceptions because of the large number of relationships thermal masses can have to one another. For example, if a dark-coloured secondary thermal mass is on the rear wall and the target is only moderately absorbant, the secondary thermal mass will receive and absorb a large percent of the insolation reflected off the target. In such a situation the full area of this rear wall cannot be counted as secondary thermal mass. The area of secondary thermal mass must not be concealed by furnishings otherwise thermal contact is lost. In such circumstances another area of secondary thermal mass must be incorporated into the design.

If the building being modelled conforms exactly to the assumptions inherent in PDP then the maximum room air temperature result will need no correction. However a change in proportional area of secondary thermal mass, or a change in thickness of that layer, will result in a different maximum room air temperature. Such corrections are possible by using the measured differences in temperature fluctuations found in Figs. 3.18 and 3.19 (Section 3.7). These differences have been used to construct a correction graph (Fig. C.2).

For example, if one models a building with a secondary thermal mass area of twice the target area (with secondary thermal mass layer thickness of 100 mm (4 in)) and the resulting room air temperature fluctuations are (from PDP) 16·7–31·1 °C (62–88 °F), from Fig. C.2 the difference between using proportional areas of 0·5 (assumed by PDP) and twice the target area is 4·1 °C (7·5 °F), thus the actual maximum room air temperature will be 27 °C (80·5 °F). As seen on Fig. 3.17, the turning point thickness of an area of secondary thermal mass twice the target area is 70 mm (2·8 in) and a thickness greater than this (i.e. 100 mm, 4 in) will have no effect on the room air temperature fluctuation.

C.7. PDP WORKSHEET

Step	Description	Value entered	Press	Display
1 CLEAR Calculator			2nd CP 2nd CMs	
2 REPARTITION Calculator		7	2nd OP 17	399·69
3 LOAD PDP DATA STORAGE PROGRAM (a) If program has not yet been keyed from listing then do this			LRN (key in as listed) LRN	
(b) Read from magnetic card			CLR (insert card)	1.
4 RECORD DATA STORAGE PROGRAM This follows step 3(a)—program is written on to side one of a magnetic card		1	2nd Write (insert card)	1.
5 LOAD DATA Data may be loaded from a magnetic card, if used before, or keyed in as follows. Where values are zero they should be entered to keep the data in sequence	*Value entered*	*Press*	*Storage Register used*	

Step	Description	Value entered	Press	Storage Register used
5—contd. Initialize		6	A	
INPUT 1	Area of unshaded south glass (m², ft²)		R/S	06
INPUT 2	Minimum acceptable room air temperature at night (from thermostat) (°C, °F)		R/S	07
INPUT 3	Average people and appliances gains per hour at night (Wh, Btu/h)		R/S	08
INPUT *4	Average outdoor temperature at night (°C, °F)		R/S	09
INPUT 5	Density of target (kg/m³, lb/ft³)		R/S	10
INPUT 6	Specific heat of target (Wh/kg°C, Btu/lb°F)		R/S	11

INPUT *7 Actual volume of noon target (m^3, ft^3)	R/S	12
INPUT *8 Area of noon target (m^2, ft^2)	R/S	13
INPUT 9 Weather skin U–value × its area (W/°C, Btu/h °F)	R/S	14
INPUT 10 Volume of air change per hour (infiltration) (m^3, ft^3)	R/S	15
INPUT *11 U–Value of floor × its area (W/°C, Btu/h °F)	R/S	16
INPUT *12 Temperature of space below (°C, °F)	R/S	17
INPUT *13 Resistance split fraction	R/S	18
INPUT *14 Morning start temperature (°C, °F)	R/S	19

* For further information see 'Notes to Worksheet', p. 194.

Step	Description	Value entered	Press	Display

5—contd.

If an error is made in any of the above entries the following
steps are performed to correct it:

If entry 7 was incorrect one enters the 'storage register used'
number — 12 — A

Entry 7 is now correctly entered — R/S
If display flashes after pressing R/S then one has entered too
many variables, i.e. one is trying to store some value into
data register 20

INPUT *15–22
Fraction of noon target area hit from hour 1 to hour 8

Hour	1	2	3	4	5	6	7	8
Value entered	20							
Press	B	R/S	R/S	R/S	R/S	R/S	R/S	R/S
Storage register used	20	25	30	35	40	45	50	55

If an error is made during entry of this table, all eight
values must be reentered: enter 20, press B and then start
with hour 1 value again. The same applies to all following
tables.
If display flashes after pressing R/S, you have tried to enter
hour 9 values!

INPUT *23–30

Horizontally directed component of solar gain from hour 1 to hour 8 (W/m², Btu/ft² h)

Hour		1	2	3	4	5	6	7	8
Value entered	22								
Press	B	R/S	R/S	R/S	R/S	R/S	R/S	R/S	R/S
Storage register used		22	27	32	37	42	47	52	57

INPUT *31–38

Vertically directed component of solar gain from hour 1 to hour 8 (W/m², Btu/ft² h).

These values are calculated within the program and placed into the following data registers:

Enter: Profile angle [   ] °

Press: C

Hour	1	2	3	4	5	6	7	8
Storage register used	21	26	31	36	41	46	51	56

INPUT *39–46

People, appliance, and other window gains from hour 1 to hour 8 (Wh, Btu/h)

Hour		1	2	3	4	5	6	7	8
Value entered	23								
Press	B	R/S	R/S	R/S	R/S	R/S	R/S	R/S	
Storage register used		23	28	33	38	43	48	53	58

Step	Description	Value entered	Press	Display
5--*contd.*	INPUT *47-54			

Outdoor temperature from hour 1 to hour 8

Hour	1	2	3	4	5	6	7	8
Value entered	24							
Press	B	R/S	R/S	R/S	R/S	R/S	R/S	R/S
Storage register used	24	29	34	39	44	49	54	59

Step	Description	Value entered	Press	Display
6 RECORD DATA	Only data registers 00–59 are needed to run PDP–data registers outside this are used for internal computation	4	2nd Write (insert inverted card)	4.
		3	2nd Write (insert card)	3.
7 LOAD PROGRAM (PDP)				
(a) If program has not been keyed from listing then do this			LRN (key in as in listing) LRN	
(b) Read from magnetic card			CLR (insert card)	1.
			CLR (insert inverted	2.

8 PROGRAM MODIFICATIONS

It is important to note that the listed positions will alter slightly when the following modifications are performed. Such changes in numbering do not affect the working of the program.

If the following table is used then the sequence is:

(Be sure to allow enough space for change—e.g. when changing 0·34 to 0·018 insert a space so that program steps are not overwritten.)

GTO
location
number
LRN
(perform
change)
LRN

Change locations towards the end of the program first and work backwards

Description	Location	Change	To	Insert or Delete
(a) Absorption	228–230	0·85		
(b) Print	217	R/S	2nd PRT	
(a) Absorption	203–230	0·85		
(a) Absorption	104–106	0·85		
(c) Louvres	054–055	0·9	1	D
(d) Metric/imperial	031–033	0·34	0·018	I
(d) Metric/imperial	021–024	3·86	0·68	

Step	Description	Value entered	Press	Display
8—contd.				
	(a) Absorption: The absorption factor within the program listing is set at 0·85 in three locations. This factor should be adjusted depending on the colour of the target area. Also if the areas outside the target are massive and dark coloured, this factor should be adjusted upward.			

Colour	Absorption factor scale
Black	0·95
Dark	0·85
Grey (concrete block)	0·70
Light	0·30

(b) Print: In order to make use of the PC–100A printer.

(c) Louvres: The program listing assumes that reflective louvres are being used. If they are not used change the 0·9 to 1.

(d) Metric/imperial: The program is written in metric units. Values in two locations must be changed in order to use imperial units.

9 RECORD PROGRAM (PDP)

Step	Description	Value entered	Press	Display
	Record program plus any modifications on to a magnetic card	1	2nd **WRITE** (insert card)	1.
		2	2nd **WRITE** (insert inverted card)	2.

10 STORE 't'

Store in 't' either 0 to model a solid target or 1 to model a liquid (isothermal) target*

	0 or 1	x ■ t

11 RUN PROGRAM

	A	Room air temp.
		Hour 1
	R/S	Hour 2
	R/S	Hour 3
	→	→
	R/S	Hour 8
	R/S	Carry-through in hours.

If the print option has been used (step 4(b)) all eight hourly room air temperatures will be printed after button A is pressed and the number of hours of carry-through will be displayed on the calculator

12 CORRECTION

Correct the maximum room air temperature achieved by the system using the correction graph (Fig. C.2)

* Validation work has not been carried out on this program to test its accuracy when modelling water as a storage medium.

Notes to Worksheet

INPUT **4**.   Average outdoor temperature at night (see Section 3.1).

INPUT **47–54**. Outdoor temperatures from hour 1 to hour 8 (see Section 3.1).

INPUT **7**.   Actual volume of noon target (see Section 3.8).

INPUT **8**.   Area of noon target (see Section 3.8).

INPUT **15–22**. Fraction of noon target area hit from hour 1 to hour 8 (see Section 3.8).

INPUT **11**.   (*U*–Value of floor × area of floor.)

INPUT **12**.   Temperature of space below.

Inputs **11** and **12** were developed for use in a situation where the target is on the ceiling, and where a cellar or enclosed unheated space is below the room being modelled. In such a situation the floor will not be conceived as being secondary thermal mass, and the temperature of the space below will be the same as the ground temperature for that time of year. If this situation is not being modelled, zero should be used for both inputs.

INPUT **13**.   Resistance split fraction (see end of Section 3.7).

INPUT **14**.   Morning start temperature.

This is the room temperature at the beginning of the first hour modelled, which is the thermostat set point.

INPUT **23–30**. Horizontally directed component of solar gains from hour 1 to hour 8 (see Section 3.8).

INPUT **31–38**. Vertically directed component of solar gains from hour 1 to hour 8 (see Section 3.8).

INPUT **39–46**. Gains from people, appliances and other windows from hour 1 to hour 8.

The gains from people and appliances vary each hour and such input can be estimated (see Fig. 3.38, Section 3.10). Gains from other windows are assumed not to fall on the target area. Being east, west or north windows it is assumed that these gains will be kept to a minimum. Such gains are assumed to go directly into heating the room air without internal program corrections for back losses.

LISTING OF : DATA STORAGE PROGRAM FOR PDP

PROGRAMMER : R.M.LEBENS   PAGE : 1

For use on a T I - 59 calculator

LOC	CODE	KEY	LOC	CODE	KEY	LOC	CODE	KEY
000	76	LBL	051	42	STO	102	01	01
001	11	A	052	03	03	103	06	6
002	42	STO	053	08	8	104	44	SUM
003	05	05	054	32	X:T	105	01	01
004	91	R/S	055	43	RCL	106	01	1
005	42	STO	056	04	04	107	44	SUM
006	03	03	057	67	EQ	108	04	04
007	76	LBL	058	34	ГX	109	08	8
008	78	Σ+	059	43	RCL	110	32	X:T
009	43	RCL	060	03	03	111	43	RCL
010	03	03	061	72	ST*	112	04	04
011	72	ST*	062	05	05	113	67	EQ
012	05	05	063	05	5	114	38	SIN
013	01	1	064	44	SUM	115	61	GTO
014	44	SUM	065	05	05	116	30	TAN
015	05	05	066	01	1	117	76	LBL
016	43	RCL	067	44	SUM	118	38	SIN
017	03	03	068	04	04	119	91	R/S
018	91	R/S	069	43	RCL			
019	42	STO	070	03	03			
020	03	03	071	91	R/S			
021	02	2	072	61	GTO			
022	00	0	073	39	COS			
023	32	X:T	074	76	LBL			
024	43	RCL	075	13	C			
025	05	05	076	30	TAN			
026	67	EQ	077	95	=			
027	34	ГX	078	42	STO			
028	61	GTO	079	03	03			
029	78	Σ+	080	25	CLR			
030	76	LBL	081	42	STO			
031	34	ГX	082	04	04			
032	43	RCL	083	02	2			
033	03	03	084	02	2			
034	94	+/-	085	42	STO			
035	34	ГX	086	01	01			
036	43	RCL	087	76	LBL			
037	03	03	088	30	TAN			
038	91	R/S	089	73	RC*			
039	76	LBL	090	01	01			
040	12	B	091	65	×			
041	42	STO	092	53	(			
042	05	05	093	01	1			
043	25	CLR	094	22	INV			
044	42	STO	095	44	SUM			
045	04	04	096	01	01			
046	43	RCL	097	54	)			
047	05	05	098	43	RCL			
048	91	R/S	099	03	03			
049	76	LBL	100	95	=			
050	39	COS	101	72	ST*			

PASSIVE SOLAR HEATING DESIGN

LISTING OF : PASSIVE DESIGN PROGRAM (PDP1)		
PROGRAMMER : R. M. LEBENS		PAGE : 1
For use on a TI - 59 calculator		

LOC	CODE	KEY	LOC	CODE	KEY	LOC	CODE	KEY
000	76	LBL	051	43	RCL	102	65	65
001	11	A	052	13	13	103	65	×
002	58	FIX	053	65	×	104	93	.
003	02	02	054	93	.	105	08	8
004	43	RCL	055	09	9	106	05	5
005	10	10	056	95	=	107	55	÷
006	65	×	057	42	STO	108	43	RCL
007	43	RCL	058	62	62	109	02	02
008	11	11	059	43	RCL	110	95	=
009	65	×	060	14	14	111	42	STO
010	43	RCL	061	85	+	112	66	66
011	12	12	062	43	RCL	113	85	+
012	95	=	063	04	04	114	43	RCL
013	42	STO	064	95	=	115	67	67
014	02	02	065	42	STO	116	95	=
015	25	CLR	066	61	61	117	42	STO
016	42	STO	067	43	RCL	118	67	67
017	01	01	068	03	03	119	01	1
018	43	RCL	069	85	+	120	67	EQ
019	13	13	070	43	RCL	121	14	D
020	65	×	071	63	63	122	43	RCL
021	03	3	072	95	=	123	67	67
022	93	.	073	42	STO	124	75	-
023	08	8	074	60	60	125	43	RCL
024	06	6	075	08	8	126	66	66
025	95	=	076	42	STO	127	85	+
026	42	STO	077	04	04	128	53	(
027	03	03	078	01	1	129	05	5
028	43	RCL	079	09	9	130	93	.
029	15	15	080	42	STO	131	03	3
030	65	×	081	63	63	132	65	×
031	93	.	082	43	RCL	133	43	RCL
032	03	3	083	19	19	134	66	66
033	04	4	084	42	STO	135	54	)
034	95	=	085	67	67	136	95	=
035	42	STO	086	76	LBL	137	42	STO
036	04	04	087	12	B	138	68	68
037	43	RCL	088	71	SBR	139	76	LBL
038	14	14	089	13	C	140	52	EE
039	85	+	090	71	SBR	141	71	SBR
040	43	RCL	091	13	C	142	13	C
041	16	16	092	42	STO	143	65	×
042	85	+	093	05	05	144	43	RCL
043	43	RCL	094	65	×	145	06	06
044	04	04	095	43	RCL	146	95	=
045	95	=	096	64	64	147	42	STO·
046	42	STO	097	65	×	148	69	69
047	63	63	098	43	RCL	149	53	(
048	43	RCL	099	62	62	150	71	SBR
049	18	18	100	95	=	151	13	C
050	65	×	101	42	STO	152	42	STO

LISTING OF : PASSIVE DESIGN PROGRAM (PDP1)

PROGRAMMER : R.M. LEBENS   PAGE : 2

For use on a TI – 59 calculator

LOC	CODE	KEY	LOC	CODE	KEY	LOC	CODE	KEY
153	66	66	204	08	8	255	65	×
154	85	+	205	05	5	256	53	(
155	43	RCL	206	54	)	257	43	RCL
156	03	03	207	65	×	258	00	00
157	65	×	208	93	.	259	75	–
158	43	RCL	209	06	6	260	43	RCL
159	68	68	210	54	)	261	17	17
160	95	=	211	55	÷	262	54	)
161	71	SBR	212	43	RCL	263	95	=
162	13	C	213	60	60	264	44	SUM
163	65	×	214	95	=	265	01	01
164	43	RCL	215	42	STO	266	97	DSZ
165	61	61	216	00	00	267	04	04
166	85	+	217	91	R/S	268	12	B
167	43	RCL	218	43	RCL	269	43	RCL
168	16	16	219	69	69	270	61	61
169	65	×	220	75	–	271	65	×
170	43	RCL	221	53	(	272	53	(
171	17	17	222	43	RCL	273	43	RCL
172	85	+	223	69	69	274	07	07
173	43	RCL	224	75	–	275	75	–
174	64	64	225	43	RCL	276	43	RCL
175	85	+	226	65	65	277	09	09
176	53	(	227	65	×	278	54	)
177	43	RCL	228	93	.	279	85	+
178	05	05	229	08	8	280	43	RCL
179	65	×	230	05	5	281	16	16
180	43	RCL	231	55	÷	282	65	×
181	13	13	232	43	RCL	283	53	(
182	75	–	233	18	18	284	43	RCL
183	43	RCL	234	54	)	285	07	07
184	65	65	235	65	×	286	75	–
185	55	÷	236	93	.	287	43	RCL
186	43	RCL	237	04	4	288	17	17
187	18	18	238	85	+	289	54	)
188	54	)	239	43	RCL	290	95	=
189	65	×	240	66	66	291	75	–
190	93	.	241	75	–	292	43	RCL
191	04	4	242	43	RCL	293	08	08
192	08	8	243	61	61	294	95	=
193	85	+	244	65	×	295	42	STO
194	43	RCL	245	53	(	296	65	65
195	65	65	246	43	RCL	297	43	RCL
196	55	÷	247	00	00	298	01	01
197	43	RCL	248	75	–	299	55	÷
198	18	18	249	73	RC*	300	43	RCL
199	65	×	250	63	63	301	65	65
200	53	(	251	54	)	302	95	=
201	01	1	252	85	+	303	42	STO
202	75	–	253	43	RCL	304	00	00
203	93	.	254	16	16	305	91	R/S

LISTING OF PASSIVE DESIGN PROGRAM (PDP 1)

PROGRAMMER : R. M. LEBENS PAGE : 3

For use on a TI – 59 calculator

LOC	CODE	KEY
306	76	LBL
307	13	C
308	42	STO
309	64	64
310	01	1
311	44	SUM
312	63	63
313	73	RC*
314	63	63
315	92	RTN
316	76	LBL
317	14	D
318	43	RCL
319	67	67
320	42	STO
321	68	68
322	61	GTO
323	52	EE

REFERENCES

1. Unpublished research by T. E. Johnson at MIT, 1975.
2. LEBENS, R. M., 'A design tool to assess room air temperatures of a passively heated space', *2nd National Passive Solar Conf. Proc.*, Philadelphia, PA, USA, 16–18, March 1978.
3. LEBENS, R. M., *Exploring Various Aspects of Passive Solar Energy Collection with Particular Reference to its Potential Use in the Rehabilitation of Nineteenth Century Row Housing in England*, MIT M.Arch. A.S. Thesis, February 1978.
3a. American Section of I.S.E.S. *Passive Systems 78*, January 1979.
4. ROGERS, B. T., 'Some performance estimates for the Wright house', *Passive Solar Heating and Cooling Conference Proceedings*, ERDA, LA–6637–C, Albuquerque, New Mexico, USA, 18–19 May 1976.
5. NORTH EAST SOLAR ENERGY CENTER, 'Validation experience: Available passive design programs', *Technical Report 2*, 70 Memorial Drive, Cambridge, MA 02142, USA, January 1979.
6. US Dept. of Commerce, *Climatic Atlas of the United States*.
7. Libbey–Owens–Ford Company, *Sun Angle Calculator®*, Merchandizing Dept., 811 Madison Ave., Toledo, Ohio 43695, USA, 1974.
8. ASHRAE, *Handbook of Fundamentals*, 1978.

Cost/Load Program

D.1. INTRODUCTION

The program may be changed from metric to imperial units by changing the average thermostat set temperature in the program listing locations (LOC) 321 to 325 to the equivalent temperature in degrees Fahrenheit. The thermostat set temperature of 18 °C was established for a dwelling in England and should be changed to 70 °F to model a dwelling in the USA. Consistency of units is necessary; for metric the following may be used: kWh, m², £($)/kWh, °C and kWh/day °C; for imperial units MBtu, ft², £($)/MBtu, °F and MBtus/day °F may be used. ·

For further details on the Cost/Load Program see Sections 3.11 and 3.18.

The following is a list of those data registers employed by the Cost/Load Program (but not for data input in the data storage program) and their purpose:

00. Counter and later holds the money saved per year by improvement.
01. Cost of heating without improvement.
02. Number of years of borrowing or life span of improvement.
03. Discount rate.
04. Fuel cost spiral rate.
05. Storage within SBRx ◪ t.
06. Dsz for looping.
07. Area of improvement (m²).
08. Cost of improvement £($)/m².
09. Cost of heating fuel.
10. Degree day base temperature or balance point temperature.
11. Heat loss per day °C(°F).

J F M A M J J A S O N D
12 13 14 15 16 17 18 19 20 21 22 23

Degrees difference and approximate heating load per month.
60. Total heating load per year.
61. The variable a in eqn (26).

62. Cost of heating per year with improvement.
63. Indirect address for degrees difference array.
64. Indirect address for mean temperature, days per month, and internal gains array.
65. Temporary storage.
66. Sum of internal gains.
67. Sum of days of internal gains.
68. Internal gains per day.
69. Cost benefit and payback period.

D.2. COST/LOAD WORKSHEET

Step	Description	Value entered	Press	Display
1 CLEAR Calculator			2nd CP 2nd CMs	
2 REPARTITION Calculator		7	2nd OP 17	399·69
3 LOAD COST/LOAD DATA STORAGE PROGRAM (a) If program has not yet been keyed from listing then do this			LRN (key in as listed) LRN	
(b) Read from magnetic card			CLR (insert card)	1.
4 RECORD COST/LOAD DATA STORAGE PROGRAM This follows Step 3(a)—program is written on to side one of a magnetic card		1	2nd Write (insert card)	1.
5 LOAD DATA Data may be loaded from a magnetic card, if used before, or keyed in as follows (where values are zero they should be entered to keep the data in sequence):				

Step	Description	Value entered	Press	Display

5—contd.

INPUT 1-12

Normal monthly mean outdoor temperatures for months January (1) to December (12) (°C, °F)

Month	J	F	M	A	M	J	J	A	S	O	N	D
Value entered Press	CLR A R/S	R/S	R/S	R/S	R/S	R/S	R/S	R/S	R/S	R/S	R/S	R/S
Display	1	2	3	4	5	6	7	8	9	10	11	12
Storage register	24	27	30	33	36	39	42	45	48	51	54	57

Note that the display will prompt the temperature entry for the month whose number is displayed (Jan. 1, to Dec. 12). Zero indicates that all monthly values have been entered.

INPUT 13-24

Number of days in each month. The number of days in each month is automatically stored by the program in data registers 25, 28, 31, 34, 37, 40, 43, 46, 49, 52, 55 and 58 by pressing CLR and then B. No entry is needed.

INPUT 25-36

Internal heat gains per month from people, appliances and useful solar energy (if no overheating then all solar energy collected is useful). kWh/month, MBtu/month

Month		J	F	M	A	M	J	J	A	S	O	N	D
Value entered													
Press	CLR C	R/S	R/S	R/S	R/S	R/S	R/S	R/S	R/S	R/S	R/S	R/S	R/S
Display	1.	2.	3.	4.	5.	6.	7.	8.	9.	10.	11.	12.	0.
Storage register used		26	29	32	35	38	41	44	47	50	53	56	59

Again the display prompts the next monthly data entry.
Zero in the display indicates that all monthly values have been entered.

If an error is made in entering a value it can either be accessed independently by entering the corrected value and pressing STO and the storage register number used, or all 12 monthly values should be reentered starting at CLR, C. This note also applies to the data Table in input 1–12 above.

INPUT 37
Heat loss per day (°C, °F)   (kWh/day °C, MBtu/day °F)   D   Heat loss/day

This is the fabric plus infiltration losses (see Section 3.11, eqn. (20))

INPUT 38
Cost of heating fuel.   (£/kWh, $/MBtu)   E   Cost of heating fuel
(i.e. Cost of useful energy supplied by the heating fuel— after the efficiency of the boiler has been accounted for.)

Step	Description	Value entered	Press	Display
6 RECORD DATA Only data registers 00–59 are needed to run Cost/Load— data registers outside this are used for internal computation		4	2nd Write (insert inverted card)	4.
		3	2nd Write (insert card)	3.
7 LOAD COST/LOAD PROGRAM (a) If program has not been keyed from listing then do this			LRN (key in as listed) LRN	
(b) Read from magnetic card			CLR (insert card)	1.
			CLR (insert inverted card)	2.
8 RECORD COST/LOAD PROGRAM This follows step 7(a)—Cost/Load is written on to the two sides of a magnetic card		1	2nd Write (insert card)	1.
		2	2nd Write (insert inverted card)	2.

9 RUN COST/LOAD PROGRAM FOR BUILDING WITHOUT IMPROVEMENTS

Instruction	Input	Key	Result
Compute balance point temperature of the building without improvements (approx. $1\frac{1}{2}$ min)		A	Balance point temperature
Enter the number of degree days for this balance point temperature (see Section 3.18, Table 3.10; or Appendix A, Table A.1) and calculate the resulting heating load per year	Degree days	R/S	Heating load without improvement
Compute the cost of this heating load (make a note of this result)		R/S	Cost of heating per year without improvement

10 ENTER DESCRIPTION OF IMPROVED BUILDING

(a) If change affects only the heat loss per day (°C, °F) then enter this and continue at 10(c)

Instruction	Input	Key	Result
	Heat loss per day	C	Heat loss per day

(b) More often than not the change will also affect the internal gains per month values. In such a case steps 3(b) and 5 will have to be repeated to enter the new values. Step 5 should be followed by step 7(b) and then this sequence:

Instruction	Input	Key	Result
Enter cost of heating per year without improvement (as noted in step 9) and continue at 10(c)	Cost	STO 01	Cost

Step	Description	Value entered	Press	Display
10—contd.				
(c)	Enter cost of improvement (£/m², $/ft²). (Register 08)	Cost	D	Cost
	Enter area of improvement (m², ft²)—if the cost entered was the total cost of the job then the area value becomes 1. (Register 07)	Area	E	Area
	Enter discount rate (fraction)—see Section 3.18. (Register 03)	Discount rate	2nd C'	Discount rate
	Enter fuel cost spiral rate (fraction)—see Section 3.18 (Register 04)	Fuel spiral rate	2nd B'	Fuel spiral rate
	Enter borrowing period or life span of improvement (y)—see Section 3.18. (Register 02)	Years	2nd D'	Years
11	RUN COST/LOAD PROGRAM FOR IMPROVED BUILDING			
	Compute the balance point temperature for the improved building (approx. 1½ min)		B	Balance point temperature
	Enter the number of degree days for that balance point temperature. (From Table 3.10, Section 3.18, or Table A.1, Appendix A) and calculate the resulting heating load	Degree days	R/S	Heating load with improvement
	Compute the cost of this heating load		R/S	Cost of heating with improvement

12 LIFE CYCLE COSTING ANALYSIS

Compute cost benefit of the improvement*	2nd E'	Cost benefit
Compute the payback period of the improvement	R/S	Payback period (y)
Any of the values entered in Step 10(c) may be altered (by entering the value and pressing the corresponding letter access key) and the cost benefit recalculated as follows:	2nd E'	New cost benefit
This may be done as many times as required to give an idea of how these difficult-to-define variables affect the payback period.	R/S	New payback period

Note:

To use only the Life Cycle Costing section of this program after loading in the program, the sequence is as follows:		
Enter savings in fuel cost during the first year	Saving	STO 00
Then continue with step 10(c) followed by:		2nd E'
		R/S

		Saving
		Cost benefit
		Payback

* This will be the payback period if the discount rate is equal to the fuel cost spiral rate.

LISTING OF : DATA STORAGE PROGRAM FOR COST/LOAD	
PROGRAMMER : R.M.LEBENS	PAGE : 1
For use on a TI-59 calculator	

LOC	CODE	KEY		LOC	CODE	KEY
000	76	LBL		051	42	STO
001	11	A		052	58	58
002	25	CLR		053	03	3
003	02	2		054	00	0
004	04	4		055	42	STO
005	42	STO		056	34	34
006	00	00		057	42	STO
007	01	1		058	40	40
008	42	STO		059	42	STO
009	01	01		060	49	49
010	76	LBL		061	42	STO
011	89	π		062	55	55
012	91	R/S		063	02	2
013	72	ST*		064	08	8
014	00	00		065	42	STO
015	03	3		066	28	28
016	44	SUM		067	91	R/S
017	00	00		068	76	LBL
018	01	1		069	13	C
019	44	SUM		070	25	CLR
020	01	01		071	02	2
021	01	1		072	06	6
022	03	3		073	42	STO
023	32	X:T		074	00	00
024	43	RCL		075	01	1
025	01	01		076	42	STO
026	67	EQ		077	01	01
027	39	COS		078	61	GTO
028	76	LBL		079	89	π
029	38	SIN		080	76	LBL
030	43	RCL		081	14	D
031	01	01		082	42	STO
032	61	GTO		083	11	11
033	89	π		084	91	R/S
034	76	LBL		085	76	LBL
035	12	B		086	15	E
036	25	CLR		087	42	STO
037	03	3		088	09	09
038	01	1		089	91	R/S
039	42	STO		090	76	LBL
040	25	25		091	39	COS
041	42	STO		092	25	CLR
042	31	31		093	42	STO
043	42	STO		094	01	01
044	37	37		095	61	GTO
045	42	STO		096	38	SIN
046	43	43				
047	42	STO				
048	46	46				
049	42	STO				
050	52	52				

LISTING OF: COST / LOAD PROGRAM		
PROGRAMMER : R. M. LEBENS		**PAGE : 1**
For use on a TI-59 calculator		

LOC	CODE	KEY
000	76	LBL
001	11	A
002	71	SBR
003	59	INT
004	43	RCL
005	60	60
006	65	×
007	43	RCL
008	09	09
009	95	=
010	42	STO
011	01	01
012	91	R/S
013	76	LBL
014	12	B
015	71	SBR
016	59	INT
017	43	RCL
018	60	60
019	65	×
020	43	RCL
021	09	09
022	95	=
023	42	STO
024	62	62
025	75	-
026	43	RCL
027	01	01
028	95	=
029	94	+/-
030	42	STO
031	00	00
032	43	RCL
033	62	62
034	91	R/S
035	76	LBL
036	10	E'
037	43	RCL
038	04	04
039	32	X:T
040	43	RCL
041	03	03
042	67	EQ
043	38	SIN
044	53	(
045	43	RCL
046	04	04
047	85	+
048	01	1
049	54	)
050	55	÷

LOC	CODE	KEY
051	53	(
052	43	RCL
053	03	03
054	85	+
055	01	1
056	54	)
057	95	=
058	42	STO
059	61	61
060	65	×
061	43	RCL
062	00	00
063	65	×
064	53	(
065	43	RCL
066	61	61
067	45	Y×
068	43	RCL
069	02	02
070	75	-
071	01	1
072	54	)
073	55	÷
074	53	(
075	43	RCL
076	61	61
077	75	-
078	01	1
079	54	)
080	95	=
081	55	÷
082	53	(
083	43	RCL
084	08	08
085	65	×
086	43	RCL
087	07	07
088	54	)
089	95	=
090	42	STO
091	69	69
092	91	R/S
093	01	1
094	32	X:T
095	43	RCL
096	69	69
097	77	GE
098	79	X̄
099	01	1
100	00	0
101	45	Y×

LOC	CODE	KEY
102	06	6
103	95	=
104	42	STO
105	69	69
106	76	LBL
107	97	DSZ
108	91	R/S
109	76	LBL
110	79	X̄
111	43	RCL
112	02	02
113	55	÷
114	43	RCL
115	69	69
116	95	=
117	61	GTO
118	97	DSZ
119	76	LBL
120	38	SIN
121	53	(
122	43	RCL
123	07	07
124	65	×
125	43	RCL
126	08	08
127	54	)
128	55	÷
129	43	RCL
130	00	00
131	95	=
132	42	STO
133	05	05
134	91	R/S
135	43	RCL
136	05	05
137	91	R/S
138	76	LBL
139	17	B'
140	42	STO
141	04	04
142	91	R/S
143	76	LBL
144	18	C'
145	42	STO
146	03	03
147	91	R/S
148	76	LBL
149	19	D'
150	42	STO
151	02	02
152	91	R/S

LISTING OF:COST / LOAD PROGRAM	
PROGRAMMER: R.M.LEBENS	PAGE:2
For use on a TI-59 calculator	

LOC	CODE	KEY		LOC	CODE	KEY		LOC	CODE	KEY
153	76	LBL		204	53	(		255	97	DSZ
154	13	C		205	91	R/S		256	06	06
155	42	STO		206	65	×		257	89	π
156	11	11		207	43	RCL		258	92	RTN
157	91	R/S		208	11	11		259	76	LBL
158	76	LBL		209	54	)		260	44	SUM
159	14	D		210	42	STO		261	71	SBR
160	42	STO		211	60	60		262	35	1/X
161	08	08		212	91	R/S		263	25	CLR
162	91	R/S		213	92	RTN		264	42	STO
163	76	LBL		214	76	LBL		265	66	66
164	15	E		215	49	PRD		266	42	STO
165	42	STO		216	71	SBR		267	67	67
166	07	07		217	35	1/X		268	76	LBL
167	91	R/S		218	25	CLR		269	33	X²
168	76	LBL		219	42	STO		270	01	1
169	59	INT		220	65	65		271	44	SUM
170	53	(		221	76	LBL		272	63	63
171	43	RCL		222	89	π		273	02	2
172	53	53		223	01	1		274	44	SUM
173	85	+		224	44	SUM		275	64	64
174	43	RCL		225	63	63		276	25	CLR
175	56	56		226	44	SUM		277	32	X:T
176	85	+		227	64	64		278	73	RC*
177	43	RCL		228	53	(		279	63	63
178	59	59		229	43	RCL		280	67	EQ
179	85	+		230	10	10		281	25	CLR
180	43	RCL		231	75	-		282	73	RC*
181	26	26		232	73	RC*		283	64	64
182	85	+		233	64	64		284	44	SUM
183	43	RCL		234	54	)		285	67	67
184	29	29		235	71	SBR		286	01	1
185	85	+		236	77	GE		287	44	SUM
186	43	RCL		237	01	1		288	64	64
187	32	32		238	44	SUM		289	73	RC*
188	54	)		239	64	64		290	64	64
189	42	STO		240	53	(		291	44	SUM
190	66	66		241	43	RCL		292	66	66
191	01	1		242	05	05		293	76	LBL
192	08	8		243	65	×		294	78	Σ+
193	02	2		244	43	RCL		295	97	DSZ
194	42	STO		245	11	11		296	06	06
195	67	67		246	65	×		297	33	X²
196	71	SBR		247	73	RC*		298	25	CLR
197	39	COS		248	64	64		299	32	X:T
198	71	SBR		249	54	)		300	43	RCL
199	49	PRD		250	72	ST*		301	66	66
200	71	SBR		251	63	63		302	67	EQ
201	44	SUM		252	01	1		303	34	ΓX
202	43	RCL		253	44	SUM		304	76	LBL
203	10	10		254	64	64		305	52	EE

LISTING OF : COST/LOAD PROGRAM	
PROGRAMMER : R.M. LEBENS	PAGE : 3
For use on a TI-59 calculator	

LOC	CODE	KEY		LOC	CODE	KEY
306	71	SBR		357	25	CLR
307	39	COS		358	32	X:T
308	92	RTN		359	43	RCL
309	76	LBL		360	05	05
310	39	COS		361	77	GE
311	53	(		362	45	Y×
312	43	RCL		363	25	CLR
313	66	66		364	42	STO
314	55	÷		365	05	05
315	43	RCL		366	76	LBL
316	67	67		367	85	+
317	54	)		368	92	RTN
318	42	STO		369	76	LBL
319	68	68		370	45	Y×
320	53	(		371	61	GTO
321	01	1		372	85	+
322	08	8		373	76	LBL
323	93	.		374	25	CLR
324	00	0		375	01	1
325	00	0		376	44	SUM
326	75	-		377	64	64
327	53	(		378	61	GTO
328	43	RCL		379	78	Σ+
329	68	68		380	76	LBL
330	55	÷		381	34	ГX
331	43	RCL		382	53	(
332	11	11		383	43	RCL
333	54	)		384	26	26
334	54	)		385	85	+
335	42	STO		386	43	RCL
336	10	10		387	59	59
337	92	RTN		388	54	)
338	76	LBL		389	42	STO
339	35	1/X		390	66	66
340	01	1		391	06	6
341	01	1		392	02	2
342	42	STO		393	42	STO
343	63	63		394	67	67
344	02	2		395	61	GTO
345	03	3		396	52	EE
346	42	STO				
347	64	64				
348	01	1				
349	02	2				
350	42	STO				
351	06	06				
352	92	RTN				
353	76	LBL				
354	77	GE				
355	42	STO				
356	05	05				

Data Bases for Worked Example

E.1. OVERHEATING PROGRAM—DATA BASES

The following data bases were collected for the solar design (Fig. 4.4 Section 4.1) for use on a program similar to the Passive Design Program (PDP) (Appendix C), but larger:

1. The area of the collection window.
Both collection rooms have a south glazed area of 4·9 m². This area does not include frames.

2. The noon target area.
On 21st October at a latitude of 52 °N the profile angle of the sun at noon is 27°.

Thus

$$p = \frac{h}{\tan P} \qquad \text{(see Section 3.8)}$$

where p = the depth of penetration of the sunlight,
   h = the height of the window,
   P = the profile angle at noon.

The noon target area is thus 10·6 m² (p × width of room).

3. The percent of noon target area hit on each of the 8 h modelled (Table E.1).

Table E.1. Percent of target hit (see Section 3.8)

	Time of day							
	09.00	10.00	11.00	12.00	13.00	14.00	15.00	16.00
Percent target area illuminated	0·25	0·42	0·73	1·0	0·73	0·42	0·25	0·15

4. The horizontal component of the energy gains for each hour modelled.
The figures in Table E.2 are obtained by reducing reference data [1] for a clear day on a vertical south-facing surface, by a dirt factor and by the

Table E.2.   The horizontal component of the solar energy gains on a clear day in October through two layers of glass and one layer of heat mirror (by Suntek)

	Time of day							
	09.00	10.00	11.00	12.00	13.00	14.00	15.00	16.00
Horizontal component (W/m²)	204	317	383	401	383	317	204	82

transmission of the transparent membrane. The transmission is found producing a graph similar to that in Fig. 3.23 (Section 3.8) for the transparent membrane used, and by finding the angle of incidence of the sun from the Sun Angle Calculator® (Fig. 3.31, Section 3.9).

5. The vertical component of the solar gains for each hour modelled (Table E.3).

These are obtained by multiplying the horizontal components (Table E.3) by the tangent of the profile angle. The profile angle for all hours of the day is taken to be that of the noon profile angle so that during program execution the total energy intake is not reduced from its true value.

Table E.3.   The vertical component of the solar energy gains per hour on a clear day in October (52° N latitude)

	Time of day							
	09.00	10.00	11.00	12.00	13.00	14.00	15.00	16.00
Vertical component (W/m²)	104	162	195	204	195	162	104	42

6. The density and specific heat of the target.

The target is considered to be concrete for the purposes of the design method calculations. This is because the design method does not model latent heat storage materials. A rule-of-thumb is then applied to the air temperature results (Section 3.7).

$$\text{Density} = 2243 \, \text{kg/m}^3$$

$$\text{Specific heat} = 0.23 \, \text{Wh/kg}\,^\circ\text{C}$$

7. The volume of the noon target area.

The target is assumed to be 150 mm thick. Therefore the volume is 1·6 m³.

8. The average U–value × the area of the weatherskin.

A detailed breakdown of the fabric losses is listed in Section E.2 of this appendix (Tables E.28 and E.29). It is seen, in Table E.29, that the heat loss per hour °C is 61·8 W.

$$\text{Area of weather skin} = 162 \cdot 12 \, \text{m}^2$$
$$\text{Average } U\text{–value} = 61 \cdot 8/162 \cdot 12 = 0 \cdot 38 \, \text{W/m}^2 \, ^\circ\text{C}$$
$$\text{Average } U\text{–value} \times \text{area} = 61 \cdot 8 \, \text{W/} ^\circ\text{C}$$

9. The volume of air change per hour.

This is $0 \cdot 93 \times$ the volume of air in the dwelling (p. 227), i.e. $0 \cdot 93 \times 258 \cdot 95$ $= 240 \cdot 8 \, \text{m}^3$.

10. U–Value and area of suspended floor.

The heat losses from the solid floor are assumed to be negligible, because it is well insulated. The U–value of the suspended floor $= 0 \cdot 16$ (Section E.2, Table E.24). The area of the suspended floor is $42 \cdot 1 \, \text{m}^2$.

11. The resistance split fraction for the target.

As in the example shown in Section 3.7, the resistance split is 0·54.

12. The hourly gains from people, appliances and other windows for each of the 8 h modelled.

  (a) Gains from people and appliances.

      The values in Table E.4 are taken from a comparison of the data shown in Fig. 3.38 and Table 3.6 (Section 3.10).

Table E.4. Hourly gains from people and appliances in a typical English dwelling

	Time of day							
	09.00	10.00	11.00	12.00	13.00	14.00	15.00	16.00
Gains (Wh)	200	200	200	1 499	1 499	200	200	200

  (b) Gains from other windows.

      The results of the solar interference boundary tests (as shown in Fig. E.2 and Table E.14 in Section E.2) are used to determine this

Table E.5. The angles of incidence of the sun at the two vertical orientations being considered (21st October, 52°N latitude), taken from the Sun Angle Calculator®

	Time of day							
	09.00	10.00	11.00	12.00	13.00	14.00	15.00	16.00
South windows	48	38	30	27	30	38	48	60
East windows	46	61	76	—	—	—	—	—

information. Clear day insolation figures are taken directly from the reference tables [1] and reduced in the same way as in Step 4 above. Tables E.5, E.6 and E.7 show the information needs and resulting values of this input data.

Table E.6. The resulting transmission values of double glazing with a layer of double sided heat mirror (by Suntek) between (21st October, 52°N latitude)

	Time of day							
	09.00	10.00	11.00	12.00	13.00	14.00	15.00	16.00
South windows	0·53	0·59	0·60	0·60	0·60	0·59	0·53	0·42
East windows	0·55	0·41	0·18	0·47*	0·47*	0·47*	0·47*	0·47*

* Values for diffuse radiation only.

Table E.7. Clear day solar gains through all windows other than the windows in the collection rooms

	Time of day							
	09.00	10.00	11.00	12.00	13.00	14.00	15.00	16.00
Solar gains (Wh)	987	1 751	2 174	2 289	1 149	856	464	200

(c) The resulting total internal gains (Table E.8).

Table E.8. The total internal gains from people, appliances and solar energy, other than the solar gains within the collection rooms

	Time of day							
	09.00	10.00	11.00	12.00	13.00	14.00	15.00	16.00
Gains (Wh)	1 187	1 951	2 374	3 788	2 648	1 055	664	399

13. The hourly outdoor temperature for each of the 8 h modelled (Table E.9).

Table E.9. Mean hourly outdoor temperatures in England in October [2]

	Time of day							
	09.00	10.00	11.00	12.00	13.00	14.00	15.00	16.00
Temperature °C	11·5	12·4	13·2	13·8	14·3	14·5	14·4	14·0

14. The morning internal start temperature.

18·3 °C

15. The temperature of the cellar.

10 °C

16. The night gains from people and appliances per hour.
This is assumed to be 293 Wh.
17. The average room air temperature at night.
The typical night-time temperature in an English dwelling is 15·6 °C.
18. The average outdoor temperature at night [2] in October.

10·5 °C

19. The hourly gains from the east window in the lower collection room
(Table E.10).
Window E4 (Fig. E.2 Section E.2) is assumed to be furnished with a solar
modulator.

Table E.10.   Hourly solar gains through window E4 (Fig. E.2)

| | Time of day | | | | | | | |
	09.00	10.00	11.00	12.00	13.00	14.00	15.00	16.00
Gains through window E4	276	247	70	57	54	50	34	19

20. The coefficients for the solar gains through the east window, E4
(Table E.11).
This coefficient is a product of the resistance split into the target and the
percent being reflected off the solar modulators. In the morning the value of
this coefficient is 0·49 (0·54 × 0·9), but in the afternoon it reduces to 0·24
because all the radiation is diffuse and only an estimated 50 % of this will be
reflected in the direction of the target.

Table E.11.   Coefficients for the solar gains through window E4 (resistance
split × percentage reflected off solar modulators)

| | Time of day | | | | | | | |
	09.00	10.00	11.00	12.00	13.00	14.00	15.00	16.00
Coefficient	0·49	0·49	0·49	0·34	0·24	0·24	0·24	0·24

21. The area and thickness of the secondary thermal mass.

This is not a requirement for the design program but it is used to correct the air temperature fluctuation results provided by the program. In the playroom the area of secondary thermal mass is $18.8\,m^2$ which is about 1.75 times the noon target area. It has an average thickness of approximately $150\,mm$. The area and volume of secondary thermal mass in the dining room will be about the same. In this room there is no end wall but the brick paved floor compensates for this.

22. The area and volume of the convective thermal mass.
Area $= 160\,m^2$, volume $= 18.94\,m^3$.

23. The specific heat and density of the convective thermal mass.
Specific heat (brick) $= 0.23\,Wh/kg\,°C$, density (brick) $= 1970\,kg/m^3$.

E.2.   COST/LOAD PROGRAM—DATA BASES

The following data bases were collected for the Cost/Load Program (Appendix D):

1. Mean daily outdoor temperature for each month (Table E.12).
2. Number of days in each month.
31st January, 28th February, etc.

Table E.12. Mean daily outdoor temperature for each month of the year at Kew, London [2]

Month	Mean daily temperature (°C)
January	4·0
February	4·9
March	6·8
April	9·4
May	12·5
June	15·9
July	16·9
August	16·5
September	14·7
October	11·8
November	7·5
December	4·9

Table E.13.   Key to the symbols used in Table E.14

Symbol	Meaning
Sun	Area in potential sunshine
Shade	Area in shade
OS	Outer south wall
IS	Inner south wall
E	East wall

3. The internal gains per month.

(a)   Solar gains

Graphic solar interference boundary tests (Section 3.4) revealed that only in December do the row houses at the opposite end of the garden obscure the sun. The sun is obscured in December for one hour at either end of the day (Fig. E.1). Figure E.2 and the information shown in Tables E.13 and E.14 provide an example of the type of data required for assessing solar gains.

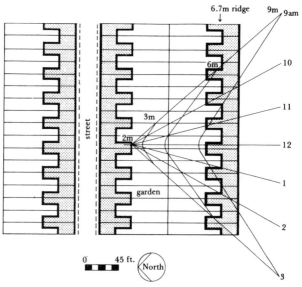

Fig. E.1.   21st December solar interference boundary graph on assumed site plan shows the sun being obscured for almost an hour either end of the day (52°N latitude).

Table E.14. Solar design. An example of the data drawn up for each month of the areas of each window in shade or in potential sunshine from combined computer and graphic solar interference boundary tests. (A similar table would be constructed for non-solar designs.) The results displayed here are for the solar design in January and November. The methods for testing are described in Section 3.4.

Window number (Fig. E.2)	Area of window(s) including frames	State	Time of day							
			09.00	10.00	11.00	12.00	13.00	14.00	15.00	16.00
OS1 North	11·6 + 1·5 = 13·1	Sun	13·1	13·1	13·1	13·1	13·1	13·1	13·1	13·1
	5·0	Shade	5·0	5·0	5·0	5·0	5·0	5·0	5·0	5·0
IS1	2·1	Sun	1·3	2·1	2·1	2·1	2·1	1·2	0·8	0·5
		Shade	0·8	0·0	0·0	0·0	0·0	0·9	1·3	1·6
IS2	3·7	Sun	0·0	1·7	3·6	3·7	0·2	1·4	0·0	0·0
		Shade	3·7	2·0	0·1	0·0	3·5	2·3	3·7	3·7
E1	1·4	Sun	1·4	1·4	1·4	1·4	1·4	1·4	1·4	1·4
		Shade	0·0	0·0	0·0	0·0	0·0	0·0	0·0	0·0
E2	1·9	Sun	0·0	1·5	1·9	1·9	1·9	1·9	1·9	1·9
		Shade	1·9	0·4	0·0	0·0	0·0	0·0	0·0	0·0
E3	2·7	Sun	1·3	2·1	2·7	2·7	2·7	2·7	2·7	2·7
		Shade	1·4	0·6	0·0	0·0	0·0	0·0	0·0	0·0
E4	2·1	Sun	1·1	1·7	2·1	2·7	2·1	2·1	2·1	2·1
		Shade	1·0	0·4	0·0	0·0	0·0	0·0	0·0	0·0

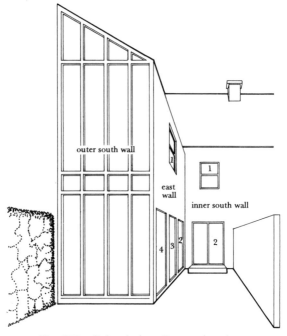

Fig. E.2. Solar design. Perspective view.

Table E.15. Clear day total insolation values (W/m^2) incident on various surfaces, taken from the IHVE guide [1]. Values are for January and November

Vertical orientation	Time of day							
	09.00	10.00	11.00	12.00	13.00	14.00	15.00	16.00
South	290	430	545	580	545	430	255	35
North	30	40	50	55	50	40	25	5
East	285	225	180	55	50	40	25	5

For the south correction factors in Table E.16, mean monthly insolation values were taken from the results of the Building Research Establishment's computer program of measured values at Kew (London) over the years 1959–1968 [3]. For the east correction factors, insolation values were taken from the UK–ISES report [4]. The methods used to determine these factors are described in Section 3.5. Both sets of correction factors were checked against the results of the method outlined in eqns. (3)–(8) (Section 3.5), using horizontal insolation values provided by Lacy [2].

Table E.16. Correction factors to convert clear day insolation values (Table E.15) to average day insolation values on the same surfaces

Vertical orientation	Month											
	J	F	M	A	M	J	J	A	S	O	N	D
South	0·29	0·29	0·45	0·49	0·64	0·70	0·60	0·56	0·51	0·44	0·37	0·34
East	0·4	0·48	0·5	0·51	0·64	0·7	0·6	0·56	0·56	0·6	0·69	0·56

Table E.17. Solar design. Average heat gain factors for a window assembly consisting of two layers of glass with a layer of double sided heat mirror (by Suntek) between for the English climate (52°N latitude)

Vertical orientation	Month											
	J	F	M	A	M	J	J	A	S	O	N	D
South	0·59	0·57	0·51	0·39	0·39*	0·39*	0·39*	0·39	0·51	0·51	0·58	0·59
East	0·45	0·55	0·55	0·55	0·56*	0·56*	0·56*	0·57	0·57	0·57	0·48	0·45

* Approximated values because these occur during the overheated months of the year.

Table E.18. Solar design. Factors determining the quantity of energy remaining within the dwelling after back-reflected losses (Section 3.8)

	Window number (Fig. E.1)							
	OS1	North	IS1	IS2	E1	E2	E3	E4
Percent of solar radiation not back-reflected	0·98	0·95	0·95	0·95	0·9	0·9	0·9	0·9

The transmission of the assembly, described in Table E.17, to normal solar radiation is 60%. The figures shown in Table E.17 are computed using the information found on Fig. 3.24 (Section 3.8), a transmission to diffuse radiation of 0·47, the horizontal insolation values provided by Lacy [2], and the method outlined in Section 3.8.

Combining the data provided in Tables E.14–E.18 the monthly solar energy gains can be computed using eqn (14) (Section 3.8). This is then multiplied by the number of days in each month and divided by 1000 to give kWh/month (Table E.19).

The solar gains through the windows of the non-solar designs (Table E.20) are computed in precisely the same way as described for the solar

Table E.19.   Total solar gains for the solar design in London (52°N latitude)

	Month											
	J	F	M	A	M	J	J	A	S	O	N	D
Solar gains (kWh/month)	294	388	671	700*	700*	700*	700*	700*	723	552	350	255

* Approximated values because these occur during the overheated months of the year.

Table E.20.   Non-solar design. Solar gains through the masonry walls (Sections 3.7 and 3.8)

	Month											
	J	F	M	A	M	J	J	A	S	O	N	D
Solar gains through the walls (kWh/month)	78	113	234	278	433	493	407	332	250	189	106	74

design above. There are also solar gains through the walls (Table E.21) which may be determined using the resistance split method described in Section 3.7.

Table E.21.   Non-solar design. Solar gains through the windows

	Month											
	J	F	M	A	M	J	J	A	S	O	N	D
Solar gains (kWh/month)	242	328	431	550	720	700*	700*	700*	449	328	201	201

* Approximated values because these occur during the overheated months of the year.

(b) Gains from people and appliances (Table E.22).

Table E.22.   Heat gains per month from inhabitants and appliances (Table 3.6 (Section 3.10)).

	Month											
	J	F	M	A	M	J	J	A	S	O	N	D
Heat gains (kWh/month)	744	672	698	600	543	450	465	543	600	698	720	744

(c) Total incidental heat gains per month (Table E.23).

Table E.23. Solar and non-solar designs. Total incidental heat gains per month from inhabitants, appliances and solar energy (combined information from Tables E.19 and E.22, and from Tables E.20, E.21 and E.22)

	Month											
	J	F	M	A	M	J	J	A	S	O	N	D
Solar design (kWh/month)	1038·4	1059·6	1368·6	1300	1242·5	1150	1165	1242·5	1322·9	1249·8	1070·4	999·1
Non-solar design (kWh/month)	1064·1	1113	1363	1428·8	1695·7	1642·8	1572·2	1574·2	1298·8	1214·3	1026	1019·5

(d) Constant heat losses to the cellar.

The cellar is assumed to be at a constant temperature of 10 °C throughout the year. Thus the heat losses to the cellar can be determined and set directly against the total incidental heat gains to give a resulting net heat gains per month which is then used in the Cost Benefit Program.

Table E.24. Solar and non-solar designs. Computation of the U–values of the suspended floor

Design	Layer	Depth (m)	Conductivity (W/°C)	Resistance (m² °C/W)	Total resistance (m² °C/W)	U–value W/m² °C
Solar	Internal air film (R_{si})			0·149		
	Carpet	0·007	0·055	0·127		
	Underlay	0·007	0·1	0·07		
	Wood	0·025	0·15	0·17	6·38	0·16
	Fibreglass insulation	0·2	0·035	5·71		
	External air film (R_{so}) ($= R_{si}$)			0·149		
Non-solar	R_{si}			0·149		
	Carpet	0·007	0·055	0·127		
	Underlay	0·007	0·1	0·07	0·665	1·5
	Wood	0·025	0·15	0·17		
	R_{so} ($= R_{si}$)			0·149		

The resulting heat losses per hour are calculated using the standard equation:

$$\text{Heat loss} = \text{Area} \times U\text{–value} \times \Delta T$$

where ΔT is the temperature difference between the two spaces, and the U–value is that detailed in Table E.24. The average indoor temperature in an English dwelling is 18 °C. Thus ΔT is 8 °C and the heat losses per month are computed. The area of suspended floor is 42·1 m².

Tables E.25 and E.26 show the resulting monthly heat losses to the cellar for the solar and non-solar designs respectively. Table E.27 takes these losses to the cellar into consideration and indicates the net heat gains for both dwelling types.

Table E.25. Solar design. Heat losses per month to the cellar (kWh/month)

	Month											
	J	F	M	A	M	J	J	A	S	O	N	D
Heat losses to cellar	40	36	40	39	40	39	40	40	39	40	39	40

Table E.26. Non-solar design. Heat losses per month to the cellar (kWh/month)

	Month											
	J	F	M	A	M	J	J	A	S	O	N	D
Heat losses to cellar	376	339	376	364	376	364	376	376	364	376	364	376

Table E.27.   Solar and non-solar designs. Resulting net heat gains (combined values of Tables E.23, E.25 and E.26) (kWh/month)

Design	Month											
	J	F	M	A	M	J	J	A	S	O	N	D
Solar	998·4	1023·6	1328·6	1261	1202·5	1111	1125	1202·5	1283·9	1209·8	1031·4	959·1
Non-solar	688·1	774	987	1064·8	1319·7	1278·8	1196·2	1198·2	934·8	838·3	662	643·5

Table E.28. Solar design. Determining the U–values of the weather skin components

Component	Layer	Thickness (m)	Conductivity (W/°C)	Resistance (m²°C/W)	Total resistance (m²°C/W)	U–Value (W/m²°C)
Roof	R_{si} (internal air film)			0·106		
	Plaster	0·016	0·50	0·032		
	Fibreglass	0·15	0·035	4·29	4·534	0·22
	R_{so} (external air film) still air			0·106		
Trash door	R_{si}			0·123		
	Wood	0·04	0·15	0·27		
	Cavity			0·18	3·776	0·26
	Expanded polystyrene	0·1	0·033	3·03		
	Wood	0·013	0·15	0·093		
	R_{so}			0·08		
Front door	R_{si}			0·123		
	Wood	0·05	0·15	0·33	0·533	1·88
	R_{so}			0·08		
Walls	R_{si}			0·123		
	Plaster	0·016	0·5	0·032		
	Brick	0·205	0·54	0·38		
	Expanded polystyrene	0·1	0·033	3·03	3·83	0·26
	Cavity			0·18		
	R_{so}			0·08		
Windows	R_{si}					
	Glass					
	Cavity					
	Double sided heat mirror		Measured U–value (Suntek)			0·85
	Cavity					
	Glass					
	R_{so}					

4. The heat loss per day °C.

(a) Fabric losses.

Fabric losses for both the solar and non-solar designs are computed using eqn (20) (Section 3.11). Reference was made to [5], [6], and [7] for the values shown in Table E.28.

Table E.29. Solar design. Determining the fabric heat loss per hour °C of the dwelling

Component	Area (m^2)	U–value (W/m^2°C)	Area × U–value
Roof	63·08	0·22	13·88
Trash door	0·63	0·26	0·16
Front door	2·24	1·88	4·21
Walls	65·85	0·27	17·78
Windows	30·32	0·85	25·77
Totals	162·12		61·8

Thus the heat loss per °C through the weather skin of the solar design dwelling is 61·8 W (Table E.29). The heat loss per day °C is determined by multiplying this by 24 (hours in the day) and dividing by 1000 to give kWh/day °C. Exactly the same process is done for the non-solar design to obtain the results shown in Table E.30. The water table is well below ground level in most parts of London [8] and thus will have no bearing on the heat losses from the solid floor of either design.

Table E.30. Non-solar design. Determining the fabric heat loss per hour °C of the dwelling

Component	Area (m^2)	U–value (W/m^2°C)	Area × U–value
Solid	20·98	0·9	18·88
Roof	63·08	0·61	38·48
Trash door	0·63	1·34	0·84
Front door	2·24	1·88	4·21
'Sandtexed' N. wall	22·36	1·63	36·45
All other walls	44·64	1·94	86·6
Windows	29·24	4·8	140·35
Totals	183·17		325·81

(b) Infiltration.

Equation (20) (Section 3.11) is used to determine infiltration losses also. See Section 3.13 ('Infiltration') for an explanation of the data bases used in Table E.31.

Table E.31. Solar and non-solar designs. The number of air changes per hour

Dwelling type	Basic air change	Inhabitant caused air change	Total
Solar design	0·18*	0·75	0·93
Non-solar design	1·5	0·75	2·25

* Attainable when dwelling unoccupied.

(c) Fabric and infiltration losses (Table E.32).

Volume of both buildings = $258·95 \, m^3$

Table E.32. Solar and non-solar designs. The resulting heat loss per day °C (kWh/day °C)

Dwelling type	Heat loss/day °C
Solar design	2·897
Non-solar design	10·562

5. Degree day table.

Table 3.10 (Section 3.18) is drawn up using the methods outlined in Section 3.11 ('Heating Load Calculations').

6. Costing (Table E.33).

The following is an outline of how the costs were broken down. Wherever possible *Spon's Architect's and Builder's Price Book* [9] was used as a reference for these figures, and updated by reference to a March 1979 Building Specification Cost Index.

The total of additional works, over and above those done in the non-solar design, is £2015·00. The value for the area of the improvement within the Cost Benefit Program will be $1 \, m^2$.

Table E.33. Costing

Component and breakdown	Cost (£)
Perimeter insulation	
Hand digging trench	
Expanded polystyrene insulation (75 mm) plus labour	101.00
Backfilling trench; depositing and compacting in layers	
Roof	
Additional 100 mm polythene faced fibreglass insulation (excluding duct). No labour charge	142·38
Roof duct	
Timber framing: worked timber plus labour	
Plasterboard and taping: cut and fixed in place	165·12
Insulation: 150 mm polythene faced fibreglass	
Fan and motor	
$\frac{1}{3}$ hp Delivers 1000 ft^3/min at $\frac{5}{8}$ in. pressure—$9\frac{1}{2}$ in. diameter wheel	47·00
Running cost for 10 y	15·00
Cutting opening through into roof/opening	5·15
Grille to opening into playroom plus labour/grille	52·50
Suspended floor	
200 mm polythene faced fibreglass insulation plus labour	204·64
North wall	
100 mm Expanded polystyrene sheets 2 × 50 mm	
Lattice timber framing in 50 × 50 studs plus labour	302·84
Expanded metal lathing plus render	
North window to utility room	
Remove window	12·00
Insert kitchen window into opening, block out in timber and plasterboard plush finish with plaster and paint including labour	100·00
East walls	
Timber framing, polystyrene insulation, expanded metal lathing and render plus labour	
South walls	547·57
Framing + insulation + lathing + render + labour and rainwater + roof extras	
North windows	
Double glazing installed	
Double sided heat mirror installed	76·00
Doors	
Weather stripping front door	
Weather stripping trash door	6·00
Insulating trash door with 100 mm expanded polystyrene	
Heating grilles	
Cutting out for grilles	25·00
Control grilles (5 No.)	62·50
Solar modulators	119·35
Saving on curtains and fixings over same windows and others; present value over 25 y.	165·00

Table E.33.—contd.

Component and breakdown	Cost (£)
PCM tiles (Sol–Ar–Tiles®)	
Cost of tiles (£15·74/m²)	
Fixing/m² (£4·00/m²)	400·00
Living-room window	
Cost of double glazed french doors with heat mirror between	398·75
Saving on non-solar lounge window	
Saving on removing wall	
Saving on quoining up	275·00
Saving on plastering	
Saving on lintel and strutting + needles and shores	
Bedroom	
Double glazing + heat mirror	50·00
E2* Double glass + double heat mirror–fixed minus proposed arrangement in non-solar design	38·00
E3* Double glazed french doors + double heat mirror minus proposed arrangement in non-solar design	67·75
E4* One layer glass + double heat mirror–fixed	50·00
E1* Storm + heat mirror	38·75
South main window	
Double special casement + double heat mirror	1000·00
Saving on dining room window	
Saving on wall removed	
Saving on quoining up	570·00
Saving on plastering	
Save on shoring + lintol + needles	
Handrail	
Worked softwood	30·00
Playroom	
Facing and plastering end of floor	15·00
Heating	
Saving on central heating system	1000·00
3·2 kWh Output gas fire fitted	65·00
Garden wall	
Remove garden wall	18·00
Quoin up	20·00

* See Fig. E.1.

7. Costing variables.
(a) Discount rate. Five percent (0·05) is taken as the test discount rate for local authority works (Section 3.18).
(b) Fuel spiral rate. Four percent was chosen as the real increase in the price of heating fuel.

(c) Cost of heating fuel. This is assumed to be £0·0114 per kWh of useful heat supplied. The estimation of the cost of North Sea Gas is fairly complex. In July 1979 the cost was £0·0114 per kWh of useful energy supplied (60 % efficiency) if costs are averaged over a year.

(d) Borrowing period. The system is assumed to have the same life-span as the building. Therefore the number of years chosen for this variable is the normal time span of a mortgage (25 y).

REFERENCES

1. IHVE, Guide A6: *Solar Data*, 1975.
2. Figures supplied by R. E. Lacy of the Building Research Establishment, computer analysis of the climate at Kew over the years between 1959 and 1968.
3. COURTNEY, R. G., *An Appraisal of Solar Water Heating in the United Kingdom*, BRE current paper CP 7/76, January 1976.
4. UK–ISES, *Solar Energy—a UK Assessment*, 1976.
5. IHVE, *Guide 1970*.
6. FAIRWEATHER, L., and SLIWA, J. A., *The Architects Journal Metric Handbook*, 3rd Edition, The Architectural Press, London, 1969.
7. Building Research Establishment, 'Standardised *U*–values', *BRE Digests: Services and Environmental Engineering*, Digest 108, Cahners Publishing Co., Boston, MA, USA, 1973.
8. Private communication with M. J. A. Pontin, Department of the Environment, Hydraulics Research Station, Wallingford, Oxon, OX10 8BA.
9. DAVIS, BELFIELD and EVEREST, Chartered Quantity Surveyors, *Spon's Architects' and Builders' Price Book*, 102nd Edition, 1977.

Index